The Farmer's Wife

The Farmer's Wife

Shares Her Life, Recipes and Tales

Authors Choice Press
New York Lincoln Shanghai

The Farmer's Wife
Shares Her Life, Recipes and Tales

Authors Choice Press
an imprint of iUniverse, Inc.

iUniverse books may be ordered through booksellers or by contacting:

iUniverse
2021 Pine Lake Road, Suite 100
Lincoln, NE 68512
www.iuniverse.com
1-800-Authors (1-800-288-4677)

Originally published by Jumbo Jacks

ISBN: 978-0-595-48965-7

Printed in the United States of America

Our Standard Abbreviations

tsp.	-	teaspoon	**sm.**	-	small
T.	-	tablespoon	**med.**	-	medium
c.	-	cup	**lg.**	-	large
oz.	-	ounce or ounces	**pt.**	-	pint
lb.	-	pound or pounds	**qt.**	-	quart
sq.	-	square	**doz.**	-	dozen
ctn.	-	carton or container	**bu.**	-	bushel
pkg.	-	package(s)	**env.**	-	envelope(s)
btl.	-	bottle(s)	**pkt.**	-	packet(s)
liter	-	liter	**mg**	-	milligram(s)
approx.	-	approximately	**gm**	-	gram(s)
temp.	-	temperature	**gal.**	-	gallon(s)

Introduction

This book is compiled with fond memories of my great-grandmothers, grandmothers, aunts, cousins and friends. Great-grandmother Airocolina P. Guidetti, grandmother Lena Ackerman, my mother Janet L. Foster, my mother-in-law Mary A. Janota, my daughter Angela, aunt Josephine Kyle, aunt Pricilla White, aunt Virginia White, my sister Jennifer L. Proper, sister-in-law Denise Proper, cousin Lori Paasch, and myself, Airocolina Chrisstina Janota, the author of this book. I dedicate it to their posterity, and to all cooks who enjoy the love of good cooking. It is with that thought in mind that I have compiled these recipes. Some of them are treasured family secrets, some are old and some are new. However, they all reflect the love of cooking and animals.

With the love of cooking also goes the love of animals. All the proceed of this book go to the Settler's Pond Hooved Animal Shelter & Rescue, in the hopes that they can save and rehabilitate many, many more animals. We are the voice for these silent, suffering animals. We have to fight for them and given them lots of very special love and attention. Some, like abused children, need special care and understanding and patience. Some will never be able to fully trust a human being again. Those are the truly heartbreaking cases. I find myself going down to the barn and sitting in their stalls, if they let me, and just talking to them.

Settler's Pond was a dream of mine that my loving husband helped me to achieve. Ten years ago we received a Hampshire baby pig that someone left in a box at the end of our driveway. I heard this shrieking coming out of this box. I trembled when I open the lid at the sight of this poor little piglet that had been the victim of some horrible torture. Someone had skinned the poor thing alive. I can't get over the Horror of it. I took her up to the barn and called the vet. At that time, he said it is just a "PIG." I said I know but she is in so much pain, can you help her. He did. He said to me I can't understand your thoughts about a barnyard animal. At that point I realized my calling. The vet, from that point until his death, always came out for us, and made sure he left pain medication for us for the animals. He never said it again to me that they were just farm animals. So many of the old time vets feel the same way, but to me, an animal is an animal, whether it be a cat or dog or one of our hooved animals.

I had to keep her clean for six weeks or until the skin healed. It was hard. She was in so much pain and, needless to say, she didn't trust me either. Through the weeks ahead I gained her trust, healed her wounds, and most of all, fell in love with this little piglet. She still lives with us now, 600 pounds later. I call her and she walks up to me and lays down. I sit on her and scratch her tummy like a dog. I go out to dinner and I don't just bring home doggie bags, but piggy bags, too.

Since that day, through word of mouth, we have been able to rescue and adopt out horses and goats. The pigs and sheep, llamas and cattle usually stay here.

Goats are so cute, especially when they are babies. They follow you around and call you maaaa. They take the bottle from you. Looking up into your eyes, melting your heart. So helpless and innocent. Then they grow up. They eat your flowers and trees, and jump on the hood of your car. We recently built a log cabin home. I always let our animals have free range. That means you could come over at any time and there might be a cow laying in the front yard with the dogs. Well, it was the day for the inspector to come out. He showed up and all the animals ran up to him with curiosity. He hesitantly got out of his car. I had to walk up to him to assure him that they weren't going to hurt him. He walked up to the cabin with me. I was standing outside with him talking about the animals and the shelter and with that one of my little goats jumped up onto the hood of his car. My eyes got so big I thought life was over as I knew it. I went up to the goat to pick him up off the hood and he danced to the other side. I went to the other side and he danced across to the other. I think I was just about to start crying and the inspector picked him up off the hood. The goat nuzzled up to the man and melted his heart. He laughed and replied "At least I brought my old car." That is how we get most of our goats. They think they're so cute, and they are, but very destructive. People bring them here because they have made pets out of them, as I have too, and can't face the thought that a lot of people eat goats, too. They say to me, please keep them for life and we do. We have been able to adopt them out as companions for older horses, and orphaned foals.

We were fortunate to receive a Wallaroo. He grew up and was adopted to a petting zoo in Florida. I go and visit him when we are down there. He has a very large pasture with trees, a pond and a little misses. We are awaiting for the arrival of our new little joey that is in intensive care right now. He is just starting to get his hair. When he does and is able to take his bottle, he is coming to stay with us also.

We are a place for abused and neglected animals or ones that just need a home. Whether people are moving or sick. Or can't care for them anymore. Sometimes, things just change in your life and you can't keep them. Please feel free to call us and we will help you help them.

Animals are my life. Dogs, cats, llamas, horses, ponies, mini horses, goats, sheep, pigs, cattle and yes, even kangaroos. The most amazing feeling I have in the world is walking out of my cabin and the animals hear me. They all call to me and run up to the fence to see what kind of goodies I have for them. They love to investigate. They are nosey and lovable, full of mischief. But most of all, in one way or another, they have been abused or neglected. Suffering in silence. I look in each of their faces, into their eyes and see the innocence and hope for unconditional love. Help me help them please. God Bless You.

Roland & Airocolina Janota
Settler's Pond Hooved Animal Shelter & Rescue
1301 E. Offner Rd.
Beecher, IL 60401

(708) 946-2448

Please call for an appointment for a tour,

This book is dedicated to the animals that we helped last year. But unfortunately some of them got here too late....

Daisy horse - passed away
Vida horse - passed away
Jersey goat - passed away
Gertrude goat - passed away
Gramma Goat and her unborn kids - passed away
Matilda goat - adopted
Penelope dog - passed away
Dinky dog - passed away
Sarah goat - adopted
Holly Belguin - passed away
Cindy Lu mini horse - adopted
Candy horse - adopted
Classy - adopted
Avalon - adopted
Dolly Llama - in rehab
Ricky Llama - in rehab
Dakota Blonde - adopted and back here
Daisy Mabel full size donkey - in rehab
And a very special heart goes out for the passing of a very special cocker spaniel.

And also Loving Memory of Mary Agnes Janota
February 19, 1922 to January 19, 2005

Dedication

To my loving husband,
Roland L. Janota
and to my children,
Angela and Eric,
and to my grandchildren,
Eva Marie and Tyler James.

A little restaurant and a
hard working man.
A lost woman working
behind the counter.
He came in for his lunch
and their eyes met.
He felt like her eyes could
see through his soul.
She could hear the kindness
in his voice.
A chance meeting and
years go by.
The two come together again.
He gathers his courage
and asks her out,
she replies softly yes.
A romance long time makes
the hearts grow together.
The love unlike no other.
Brings the two together
for years to come.

~~~~~~~~~~~~~~~

A farmer's life is
full of strife
"Oh what I'd
do for a
muscular
wife."

Notes &
Recipes

Table of Contents

FAVORITE RECIPES
FROM MY COOKBOOK

Recipe Name	Page Number

Main Courses
& Meats

New arrival at Settler's Pond. We lay with them and let them feel our love.

Belguin baby passed 4 days old

A beautiful young
woman in nurse's training.
A roommate two best friends.
A construction worker and
his best friend. A party
and a chance meeting.
The two young men and
the two young nurses to be.
The young man they
called the "Milkman"
approached the
beautiful brunette nurse.
They called him
that because he always
wore white. Her eyes
taking quick glances at
him as he approached.
The feeling in her
tummy like butterflies
that she had never
experienced before.
When the walk across
the room (that seemed
like it was 10 miles)
came to an end, the
Milkman asked her,
What is your name?
She meagerly replied
Janet. A glance and a
touch of hearts and a
romance that grew to
more than love.

Main Courses & Meats

Meat Sauce Over Potatoes

2 lb. ground beef
3 lg. cans tomato paste
2 lg. cans water
1 c. sugar

Salt & pepper, to taste
2 sticks of butter
Mashed potatoes

Brown the hamburger and drain off the grease. Put 2 sticks of butter into large pot and melt. Add 1 small onion, chopped. Cook until clear. Add tomato paste, water, sugar, salt and pepper. Stir constantly. Will bubble and spit all over the stove. Add hamburger and heat back up. Spoon over mashed potatoes.
I like to serve it with garlic toast.

Easy Hors d'oeuvres

Cream cheese
1/4 c. milk

2 pkg. chipped beef
Green onions

Mix the cream cheese and milk. Beat until smooth and spreadable. Spread a spoonful onto a piece of chipped beef. Place an onion on one end of the meat and roll up around the onion. Store in refrigerator until serving.

Aunt Pricilla's Lasagna

1 lb. lasagna noodles
1 lb. sweet Italian sausage
1 lb. hamburger
2 lg. cans tomatoes
3 cans tomato sauce
Salt & pepper, to taste
1/2 tsp. oregano
1/2 tsp. parsley
1 bay leaf

2 tsp. sugar
1/2 stick butter
1 tsp. cilantro
2 to 3 cloves garlic, chopped fine
Swiss cheese
American cheese
Mozzarella cheese
Cottage cheese
Parmesan cheese

Cook pasta as directed on the box. Brown the Italian sausage and hamburger. Drain off grease and add remaining ingredients, except cheeses. Put a layer of sauce on the bottom of the pan and a layer of noodles. Layer a slice of each of the cheeses, then some cottage cheese, more sauce , then noodles, etc. Continue until pan is full. Bake at 350° until done.

Spaghetti Sauce

Neck bones
Hamburger
2 lg. cans tomatoes
3 cans tomato sauce
Salt & pepper, to taste

1/2 c. parsley
1 bay leaf
1/2 stick butter
1 tsp. cilantro
2 to 3 cloves garlic

Brown hamburger and drain off the grease. Mix remaining ingredients and simmer on low, taking note to not let it burn. Boil your spaghetti and drain it off and put on plate. Spoon your sauce on it and sprinkle with Parmesan cheese, if desired.

Stewed Duck

1 lb. bacon
1 duck, cut up
1 onion
1 celery stalk

Parsley
1 bay leaf
Mushrooms

Fry the bacon and cut up. Cut up the duck and place in bacon grease. Brown duck until crispy. Place the duck and drippings in large Pyrex oven-safe bowl. Cut up onion and celery stalk. Cover duck with water and bake for 2 hours at 350°. In the frying pan that you fried the bacon in and the duck, put 1/2 stick margarine and 2 tablespoons flour. Cook until thick and stir into juice after the duck is removed to make a gravy.

Christmas Shrimp Tree

1 medium Styro Cone form	Kale (curly lettuce)
1 (3/4") dowel rod	Cherry tomatoes
Plaster of Paris	Star fruit
Large pot	Large hors d'ouerve toothpicks
Large shrimp	

Cut dowel rod to 18 inches. Mix plaster of Paris to directions on box. Pour in large flower pot, let firm up a bit and insert dowel rod into the center; let harden. Place Styro cone form onto the standing dowel rod to form a tree. Take your kale and start at the bottom of the styro and work your way up, using frilly toothpicks with the kale, covering the styro cone.

When it is completely covered, take your cleaned peeled deveined shrimp and place them like ornaments on the tree. Do the same with the cherry tomatoes. Cover the entire tree like you are decorating your Christmas tree. Slice a piece of star fruit and place it on the top of the tree.

I like to cover the pot with a Christmas cloth napkin and I place a bowl of shrimp cocktail sauce and lemon slices around the base. Be sure you use a flower pot large enough to weight down the top-heavy tree. Cover with plastic wrap and bring out right before you serve. It makes a grand entrance and a beautiful centerpiece.

Easy Shrimp Dip

2 pkg. cream cheese	1 lb. cleaned, deveined, cooked
1/4 c. milk	shrimp
1 jar cocktail sauce	

Cream cream cheese and milk together until well mixed. Spread out onto decorative platter and pour cocktail sauce on top of the cream cheese mixture and spread lightly. Make sure shrimp are thawed and dry off with paper towel. Place them on top of cocktail sauce. Serve with Ritz crackers. Refrigerate when not eating.

Sausage, Burger Foos

1 lb. Jimmy Dean sausage	1 lb. Velveeta cheese
1 lb. ground beef	1 T. Worcestershire sauce

Brown sausage and hamburger; stir in cheese and melt. Mix in Worcestershire sauce. Place spoonful onto rye bread slices found by the deli counter. Place on cookie sheet, and either freeze then put in plastic bag and take out as you need, or bake at 350° until just turning brown.

Open-Faced Hamburgers

2 lb. ground beef
1 sm. onion, diced
2 eggs

1 c. bread crumbs
Salt & pepper, to taste
Catsup

Mix all ingredients with the ground beef, except ketchup. Open your hamburger buns and place on a cookie sheet. Gently lightly press meat into the buns, but do not make a patty. Leave a little openness to the meat, if you can understand me. Drizzle catsup on top of each one and broil for about 9 to 13 minutes.

The juice goes into the bun and adds a considerable amount of flavor.

Meat Loaf

2 lb. ground beef
2 eggs
1 1/2 c. bread crumbs

1/2 c. catsup
Salt & pepper, to taste

Mix ground beef and eggs in large bowl with your fingers. Add bread crumbs and salt and pepper, to taste. Now add the catsup and mix well with your hands. Put in a Pyrex baking dish and pat into a loaf shape. I like to spread additional ketchup on the top and bake at 350° until done, about 45 minutes to 1 hour, depending on the shape and thickness of your loaf.

Cocktail Meat Balls

2 lb. ground beef
2 eggs
1 sm. onion, diced very fine

Salt & pepper, to taste
1 c. seasoned bread crumbs

SAUCE:
1 jar cocktail sauce
1 med. jar grape jelly
Salt & pepper, to taste

2 T. Worcestershire sauce
1 T. mustard seed

Mix burger and other ingredients. Shape into meat balls and fry. Drain off grease and, in separate saucepan, combine cocktail sauce and jelly; mix well and cook over low heat until melted. Add remaining ingredients and simmer about 20 minutes. Pour over meat balls.

I like to put them in a chafing dish to keep warm. Put some frilly toothpicks out by them and plates.

Mmm good.

Beef Jerky

Flank steak
1 c. soy sauce
1 c. brown sugar

1 tsp. liquid smoke
Salt & pepper, to taste
1/2 tsp. cayenne pepper

Slice flank steak into thin strips. In a large bowl that can be seated, mix all ingredients and add milk. Soak meat in the solution 24 hours. Drain off liquid and place in food dehydrator until done, about 145° to 190°.

Quail Pie

Quail
Cooked pasta
1/2 c. water
Salt & pepper
Parsley

Butter
Store-bought pie crust
2 T. flour
1/2 stick butter

Cut up quail and place in frying pan. Salt & pepper meat and brown in butter. After it has browned and crisped up a bit, take the pasta and cook per instructions. Place in bottom of pie plate. Take the drippings from the frying pan and the 1/2 cup water and 2 tablespoons flour; cook mixture in frypan until gravy is achieved. Pour over quail. Dot with butter and put the pie crust on top of the quail. Bake at 350° until browned.

Prairie Chicken

1 lg. prairie chicken
Butter

Salt
Pepper

Boil chicken until tender. Take out of the water and cool slightly. Run with butter and salt and pepper, to taste. Place in a baking dish and put a butter pat on the breast. Bake until golden brown.

Chicken and Wine

8 boneless, skinless chicken breasts
8 slices Swiss cheese
1 can cream of chicken soup

1/4 c. white cooking wine
1 box Stove Top stuffing
1 1/2 sticks butter, melted

Grease 9x13-inch pan; put chicken in pan. Cover each piece with cheese. Mix wine and soup; pour over chicken and cheese. Sprinkle stuffing over the top and bake, uncovered, at 350° for 55 minutes. Yield: 8 servings.

Chili Soup To Can

8 qt. tomato juice
10 lb. hamburger, browned
3/4 c. onion, chopped
1 gal. navy beans
1 gal. kidney beans

3/4 c. salt, or to taste
2 lb. brown sugar
1 tsp. red pepper
1 T. chili powder

Mix all together and cook for 1 minute. Mix water and 1 quart clear jell; add to soup, stirring constantly. Boil 5 minutes, then put in jars. Cold pack for 3 hours, or pressure for 35 minutes on 10 pounds. When you open a jar, add water to make right consistency.

Stuffed Pork Chops

Pork chops, 2" thick
2 c. bread, cubed
1/4 c. melted butter
1 tsp. sage

1/4 c. chicken broth
Salt & pepper
Shredded American cheese

Slice pork chops almost in half like you're going to make 2 cups out of one. Leave 1 inch to back. Stuff with stuffing. Brown in skillet on both sides. Place in baking dish, with a little broth in bottom of dish. Cover and bake at 350° for 1 1/2 hours.

For stuffing, toss bread and melted butter. Salt and pepper, to taste. Sprinkle sage on top. Add broth and mix well.

I stick toothpicks in pork to keep stuffing in.

Ham Glaze

1 c. brown sugar
1/2 tsp. dry mustard
1/2 tsp. cinnamon

1/2 tsp. nutmeg
4 T. pineapple juice

Cook over low heat until dissolved. Pour over ham and cook, uncovered, until glazed.

Sweet and Sour Pork or Chicken

2 lb. lean pork, cut in 2"x2 1/2" strips
2 T. grease
4 1/2 oz. pineapple chunks
1/4 c. brown sugar
2 T. cornstarch
1/4 c. vinegar

2 T. soy sauce
1 green pepper, cut in strips
1/4 c. sliced onion
2 T. cherry juice
1 sm. can water chestnuts

Brown meat strips. Combine rest of ingredients and cook on low for 2 to 3 minutes, or until thick. Peppers and onions will be crisp. Serve over white rice.

Swiss Steak

1/4 c. flour
2 lb. round steak, cut in 1" cubes
3 T. grease
1/2 c. chopped onion

1 lg. can tomatoes
2 T. chopped green pepper
Salt & pepper, to taste

Brown cubed meat; add 3 tablespoons grease (Crisco) and flour to meat and drippings. Simmer and add remaining ingredients. Transfer to a 9x13-inch pan. Cook, covered, 1 1/2 hours at 350°; uncover and cook 15 minutes more.

Beef Stew

2 lb. cubed beef
1 tsp. Worcestershire sauce
1 clove garlic
1 onion, diced
Bay leaf
Salt & pepper, to taste

1 tsp. sugar
1/2 tsp. paprika
6 carrots, chopped
6 potatoes, chopped
2 cans gravy
2 beef broth cubes

Brown beef and keep juice in pan. Add remaining ingredients. Simmer until tender. To thicken, mix 2 tablespoons cornstarch into 1/4 cup water; stir and gradually add to thicken and stir. Cook a bit after adding cornstarch.

White Castle-Style Burgers

1 lb. hamburger
2 cans cream of mushroom soup

6 oz. Cheddar cheese
Hamburger buns

Brown the hamburger; drain off grease. Add 2 cans mushroom soup to your burger. Add 6 ounces shredded cheese. Simmer on low until cheese melts. Serve on buns.

Shepherd's Pie

Leftover pork roast
Mashed potatoes

Gravy
American cheese

In a deep, glass baking dish, put 1 layer of 5 butter pats. Cube pork and sprinkle on top of butter. Pour gravy on top of the mashed potatoes. Continue layering until the pan is 3/4 full. Top with gravy and slices of American cheese. Cover and bake at 350° for 25 to 35 minutes, or until hot and bubbly.

Country-Fried Steak

6 med. cube steaks
1 med. onion, sliced
1 can cream of mushroom soup

1/2 c. milk
Salt & pepper
Water

Flour each cube steak. Fry until browned; salt and pepper to taste. Put in baking dish. Mix milk and soup together and pour over steaks. Layer with sliced onions and almost cover with water. Bake at 325° for almost 2 hours, until tender.

I also pound and tenderize round steak.

Italian Roast Beef

5 1/2 lb. rump roast
1/2 tsp. salt
1/2 tsp. pepper
1/2 tsp. Italian seasoning
1/2 tsp. oregano

1/2 tsp. garlic salt
3 lg. onions
1/2 tsp. seasoned salt
1/4 tsp. basil
1/2 tsp. onion salt

Put roast in roaster; add onion and salt. Fill roaster halfway full of water. Cover and bake at 300° until tender, 3 to 4 hours, turning over every 1/2 hour. Cool and refrigerate overnight in broth.

The next day, scrape off fat from broth. Remove meat and slice very thin. Strain liquid and add remaining ingredients. Pour over sliced beef. Heat only until hot.

Brine For Meat

1 1/2 c. sugar
1 c. Tender Quick
2 T. liquid soap

2 gal. water
1 1/2 tsp. sugar cure spice

Stir until sugar is dissolved. Cover meat with brine and soak for 48 hours. Drain and roast meat.

This makes chicken taste like smoked turkey. I especially like it on wild meats, as it helps to tenderize the meat and gives a cured flavor.

Brine For Curing Ham and Bacon

1 1/2 c. Tender Quick
2 1/2 c. brown sugar
1 T. black pepper

6 c. water
1/4 c. liquid smoke
1 c. maple syrup

Soak whole hams 4 days; if sliced, soak 16 to 22 hours. For bacon, if whole, 24 hours; if sliced, 6 hours. Wrap and freeze. Hang in smokehouse and hot smoke. Hickory chips work best for great flavor. Smoke until done, about 18 to 24 hours.

Brine To Cure Meat

10 c. water
1 c. Tender Quick
1 c. packed brown sugar

2 T. liquid smoke
1/2 tsp. saltpeter

Soak pork or turkey in this brine for 3 days, then drain off. Bake meat until done or freeze until you bake.

"For Grilling" Deer Steaks

1/2 c. water
1 c. vegetable oil
1 sm. btl. soy sauce

1 clove garlic, smashed
1 tsp. parsley
1 T. coarse ground pepper

Marinate deer meat in the refrigerator overnight. Drain off and put on hot grill.

Bologna To Can

40 lb. turkey or chicken
30 lb. lean hamburger
2 lb. Tender Quick
2 lb. brown sugar
3 T. black pepper

2 3/4 T. saltpeter
2 T. garlic powder
11 T. liquid smoke
1 gal. water

Grind first 3 ingredients twice. Let stand 24 hours. Add 1 gallon water to remaining ingredients and mix well. Grind 1 more time. Pour into jars and can 3 hours, or stuff in 4-inch bags and bake on low heat at 250° for 3 1/2 hours, or until internal temperature is 160°. Also, put a small pan of water in oven. The bologna will be more moist.

Chicken Bologna

25 lb. raw, deboned chicken
3/4 lb. Tender Quick
1 oz. black pepper

2 tsp. saltpeter
2 tsp. garlic (opt.)
2 1/2 T. liquid smoke

Grind chicken and Tender Quick twice. Let stand 24 hours. Add remaining ingredients and grind again. Mix well. Put in jars and process.

Deer Bologna

40 lb. meat
10 lb. bacon
1 3/4 lb. Tender Quick
1 c. sugar
1 1/2 c. brown sugar

1/2 c. black pepper
1/4 c. Lawry's salt
1/4 c. Worcestershire sauce
5 oz. liquid smoke
1 (20 oz.) loaf bread

Grind everything together, except the bread. Put in refrigerator 4 days. Grind meat again. Soak the bread in water. Squeeze it out of the bread and add to the meat as you grind it. Mix thoroughly and pack in jars. Process as for any fresh meat.

Note: If you pack this into bags and cook in water, add 1/3 cup salt to the above ingredients.

Corned Beef

BRINE:
1/2 lb. baking soda
1/4 lb. saltpeter
2 lb. brown sugar
2 T. liquid smoke

Water (enough to cover meat well)
50 lb. chunk beef
3 qt. salt

Place meat in large crock; add salt gradually. Let stand overnight.
Rinse off and return to crock. Pour brine over meat. This will be cured and ready to use in 2 weeks. It can be canned or frozen. If the crock is kept in a cool place, the meat may be kept in brine and used within 3 months. Wash off meat before using (cold pack 3 hours).

Salami

2 lb. hamburger
2 T. Tender Quick
1 1/2 tsp. liquid smoke

1/2 tsp. garlic salt
1/2 tsp. pepper
1 c. water

Mix and chill for 12 hours. Form into loaves and bake 2 hours at 325° (when formed with bread pans), less if smaller pans.

Deer Sausage

35 lb. deer meat
6 lb. bacon or bacon ends & pieces

5 oz. sausage seasoning (more, if
    desired)

Mix all ingredients together, then grind 2 times. Stuff into casings. Freeze or make balls into jars and add the drippings, along with a little water. Process as you would any other meat. May also be put raw in freezer.

Sausage

30 lb. meat
1 1/2 c. brown sugar
1 c. flour
6 T. salt
2 1/2 T. black pepper

6 T. sage
1 tsp. nutmeg
2 1/2 T. red pepper
1 qt. water

Mix all ingredients; add to meat. Gadually add water. Press in bread loaf pans. Bake at 350° until done. Refrigerate unused portions.

Oldtimers

3 T. flour
1 c. cornmeal
1 c. milk
2 tsp. sugar

1/2 tsp. salt
2 3/4 c. boiling water
8 oz. bulk pork sausage, cooked,
    drained & crumbled

In saucepan, put flour and cornmeal; mix. Add milk, sugar, salt and sausage. Gradually stir in water. Cook and stir until thickened and bubbly. Reduce heat and cook 8 to 10 minutes more, or until very thick, stirring occasionally. Cool a bit. Put into loaf pans and cover with plastic wrap; refrigerate overnight. To serve, unmold and cut into 1/3- to 1/2-inch slices. Dip both sides into flour and fry until brown. Serve with maple syrup.

BBQ Sauce

1 c. catsup
1 T. salt
Pepper
3/4 c. brown sugar
2 T. Tabasco sauce

1/2 tsp. onion powder
4 T. Worcestershire sauce
2 T. liquid smoke
2 T. mustard seed
3 T. butter

Bring to a boil and boil until thickens. Pour into jars and refrigerate.

Stromboli Chicago Style

Sweet Italian sausage
Hard salami
Ham

Canadian bacon
Mozzarella cheese
Pizza sauce

CRUST:
3 c. flour
1/2 tsp. salt
1/4 c. oil

3/4 c. milk
1/2 c. lukewarm water
2 pkg. yeast

Chop and fry all of the meats; set aside. Mix the first 4 ingredients well. Mix 2 packages yeast with lukewarm water. Incorporate with flour mixture. Grease well a cookie sheet. Press out the dough to almost the full size of the cookie sheet. On half of the crust, spread sauce, meats and cheese. Fold over the crust and seal the edges. Put in preheated oven at 350° until done.

Stromboli

Sweet Italian sausage
Sliced pepperoni
1 can pizza sauce
Sliced mushrooms

Diced onions
Green peppers
Mozzarella cheese

CRUST:
3 c. flour
1/2 tsp. salt
1/4 c. oil

3/4 c. milk
1/2 c. lukewarm water
2 pkg. yeast

I usually fry 3 pounds sweet Italian sausage. Slice your pepperoni and set aside. Mix flour, salt, oil and milk. Dissolve the yeast in 1/2 cup lukewarm water. Mix well to form a dough. Press dough out on well-greased cookie sheet.

I like to use oblong pans. Spread 1 side with pizza sauce; sprinkle with sausage, pepperoni and cheese. Add any other toppings you want. Fold over in half and press the edges to seal. Bake at 350° until brown.

Breaded Pork Chops

Boneless thick-cut pork chops
Eggs
Cracker crumbs

Salt
Pepper

Beat the eggs and dip pork chop into and place into cracker crumbs. Press into chop. Place oil in frying pan. Put chops in frypan when oil is hot. Brown both sides. Sprinkle with salt and pepper, to taste. Put on cookie sheet and bake at 350° until done. Depends on the thickness of the chop on low long to bake them.

Eggplant, Portobello Parmesan

1 lg. eggplant	Eggs
Portobello mushrooms	Salt & pepper, to taste
Cracker crumbs	Garlic bread

SAUCE:

Tomato paste	1 tsp. Italian seasoning
1 stick butter	Fried sweet Italian sausage
2 T. sugar	Sprinkle with Parmesan cheese
1 tsp. oregano	

Fry 1 pound Italian sausage. Drain off grease; put sausage aside for sauce. Pour vegetable oil in pan, enough to fry eggplant and mushroom. While that is heating, slice your eggplant. Beat an egg and dip eggplant in egg. Put it into your cracker crumbs and press crumbs into eggplant. Set into hot oil in frypan. Fry both sides and set aside.

Clean your mushrooms and fry in butter, enough to just turn the mushroom light brown.

Sauce: Put paste and butter in saucepan. Melt butter and add seasonings. Put sausage into your tomato sauce. Simmer. Add 1/2 cup water if too thick. Slice and butter both sides of thick-cut French bread. Sprinkle with garlic salt. When they have lightly browned under broiler, place 1 slice of fried mushroom on top. Then a slice of your breaded and fried eggplant. Spoon sauce over the top and sprinkle with Mozzarella cheese. Place under broiler until cheese is melted and bubbly.

Everything is already cooked, you just have to warm it through.

Chipped Beef Dip

1 pkg. cream cheese	2 T. milk
1 pkg. chipped beef	1/4 c. minced onion

Cream softened cream cheese and milk to smooth. Add minced onion and mix well. The chipped beef is the lunch-meat type, sliced very thin. Chop in small pieces. Mix in with the cream cheese and put in decorative bowl. Refrigerate 1 hour before serving.

It serves well with pretzels.

Chili

1 1/2 lb. ground beef, browned
4 lg. cans diced tomatoes
3 T. chili powder
1/2 tsp. onion powder
1/2 tsp. garlic powder

1 tsp. hot chili powder
4 T. sugar
4 lg. cans kidney beans
Hot pepper (opt.)

Brown the ground beef and drain off the grease. In a large pot, pour the tomatoes and spices. Bring up to a boil and simmer for 1 hour. Now, add the kidney beans, heat through and serve.

Notes &
Recipes

Pasta, Noodles & Casseroles

Beauty and the Beast

One of our baby kangaroos

Horror

What a horrible place. As far as the eye can see cages, cages, cages. This man, this frail sneaky little man. He was in trouble before. The men and women came in their uniforms and shiny badges and took all of the animals before. He thinks that no one will know. But he is forever being watched.

We never had much to eat. Sometimes we would go for days without food. Water, when we got it, was put in our green slimy bowls. We would lap it up as fast as he would pour it. He would hit at the cages to scare us back from drinking when he was pouring, he said that if we aren't letting him get back to his life fast enough.

Today was a good day. We got some bread. I almost felt like I was full. The next day I saw that thin frail man walking toward a cage below me. There was a terrible smell coming. He stopped in front of my cage and his eyes glared at me. He reached in below me and took out Trixy. He threw her limp lifeless body on the ground. He cut her in pieces and threw her in the cages. The other animals didn't know that it was Trixy, my neighbor, my friend between the cages. They began to devour her. Soon he opened my cage and threw in a leg. I just sat there. How can this be happening? Please someone help me. Help us. He had run out of food. He was throwing in the dead carcass to us for food. It was either eat or eventually be eaten.

The heat of the day in August seemed to scorch us. It was particularly hot with no breeze. Please GOD don't let me live like this, please take us. Take me now. There was a commotion at the front of the house. I could hear loud horns and people. Oh, could this be the day? I shook in my cage. Waiting. I tried to see but there were too many cages.

All of a sudden there was this face, an unfamiliar face. She looked in my eyes. What was this, who was this? She opened my cage and I ran to the back. I was so afraid, but I also felt like she would be my angel. I very carefully went to the tip of her fingers and smelled. She smelled sweet. I leaped into her arms. It felt strange yet glorious at the same time. I was happy, sad, scared, and oh, so glorious. At the same time. I liked her face. This was the first time I have ever seen the back side of my cage. There were many people, men and women, carrying cages and taking my friends out one by one.

There were so many of us it took them days to empty out the cages. We went to this place. It smelled strange, very clean. There were people with white jackets on. They tended to our wounds, they gave us baths and removed the terrible fleas that took so many of my friends. I then was put in another cage. Soon there were people again with those walking cages. They loaded us back up. I was in the back of a van sitting alongside Buster, he was a large dog, lots of hair, he made me laugh.

We were in the van for a long time. The smells were strange, lots of smells. I could not pick out just one. We turned off the hard road onto a gravel one. The smells were so strong. It was so exciting. We stopped and the van doors opened. Buster and I were delivered to Settler's Pond Hooved Animal Shelter & Rescue. They let us out of the cages. I looked around and saw many strange animals. They were all talking to me. They all said welcome. You will like it here. I ran and ran and then I turned around and ran back. Buster and I were so happy. Oh what is this? There were more dogs, and cats, everyone was happy and all getting along. Horses ran up to the fence to greet us. I was so happy. Today is the day Buster and I were Born-over. It is a wonderful place here. Food, water, trees and a pond for cooling off in the summer. A place to go in the cold of winter. I love it here. Everyone loves it here. Thank God.

Pasta, Noodles & Casseroles

Chicken Broccoli Casserole

1 can cream of mushroom soup
3/4 c. milk
2 c. diced cooked chicken

1 (10 oz.) pkg. frozen broccoli
1 box chicken stuffing

Prepare stuffing as directed on box. Mix soup and milk; pour into 9-inch square baking dish (pan). Layer chicken over soup; broccoli should be layered over the chicken. Top with the stuffing. Bake 1 hour at 350°.

One-Dish Pizza Pie

1 lb. sausage
2/3 c. pizza sauce
8 slices bread
8 slices cheese

2 c. milk
3 eggs
1/2 tsp. salt

Fry meat. Mix in pizza sauce. In baking dish, layer 4 slices of bread, then meat mixture, 4 slices of cheese and 4 slices of bread. Beat together eggs, milk and salt. Pour over top. Let stand 6 hours, overnight. Bake at 350° for 45 minutes.

Scalloped Potatoes and Ham

5 lb. potatoes
2 1/2 lb. ham
4 c. milk

6 T. flour
3 T. butter
1/2 lb. Velveeta cheese

Boil potatoes in skins; peel and slice when cooled. Chop ham. Mix flour, milk and butter together and heat. Add cheese to milk mixture. Mix all together and put into large casserole dish. Bake at 250° for 1 1/2 to 2 hours. Yield: 20 servings, or 4 lumberjacks.

Beef Stroganoff

1 sm. onion, minced
1 clove garlic, minced
1/2 c. butter
1 1/2 lb. ground beef
2 T. flour
Salt & pepper, to taste

1 (8 oz.) can sliced mushrooms
1 can cream of chicken soup
1 c. sour cream
2 T. parsley, minced
Cooked noodles

Sauté onions and garlic in butter; add ground beef and brown well. Add flour and cook a bit more. Salt and pepper to taste. Mix soup and sour cream together; add mushrooms and parsley. Mix well with the hamburger. Cook until hot and pour over cooked noodles.

Hamburger and Potato Casserole

1/2 c. chopped onion
1/2 c. chopped green peppers
2 T. Crisco
2 lb. hamburger
1 c. milk

1 can cream of mushroom soup
Salt & pepper, to taste
2 c. shredded raw potatoes
Bread crumbs, for topping

Sauté onions and green peppers in Crisco. Add hamburger and fry until done; crumble well. Add milk and soup and stir well. Alternate hamburger mix, shredded potatoes and salt and pepper in well-greased, 2 1/2-quart casserole. Finish with layer of burger. Cover with bread crumbs and bake at 375° for 1 hour.

Beef and Cheese Casserole

1 lb. ground beef
1 1/2 c. cooked spaghetti
1 c. tomato juice
3 T. flour
3/4 c. cheese

1 sm. onion (opt.)
2 T. butter
2 c. milk
1/2 c. cheese (topping)

Cook spaghetti and drain off. In frypan, brown beef, butter and onion. Add flour, seasonings and milk to frypan. Cook until thick. Mix spaghetti with cheese. Place 1/2 spaghetti mixture in greased 9x13-inch pan. Pour meat mixture on top, then more spaghetti. Pour 1 small can tomato sauce over spaghetti, then meat mixture. Sprinkle with 1/2 cup cheese and bake at 350° for 25 to 30 minutes.

I use Cheddar or Mozzarella cheese; use what you like.

Brunch Casserole

8 slices of bread
4 eggs
Salt & pepper, to taste

1/2 lb. grated cheese
1 1/2 lb. cooked breakfast sausage
2 c. milk

Grease 9x13-inch pan. Cube bread and sprinkle in bottom of pan. Top with sausage, then cheese. Mix together eggs, milk, salt and pepper, to taste. Pour over sausage and cheese. Bake at 350° for 45 minutes to 1 hour. This may be made the night before and stored in the refrigerator.

Bacon Quiche

14 slices bacon, fried & crumbled
1 c. shredded Cheddar cheese
1/2 c. chopped onion (opt.)
2 c. milk
1 c. Bisquick

4 eggs
Salt & pepper
Chopped green peppers (opt.)
Mushrooms (opt.)

Pam a 10-inch pie plate. Sprinkle bacon, optional vegetables and cheese into pie plate. Beat milk, eggs and Bisquick together for about 1 minute. Pour over bacon in pie plate. Bake at 400° for about 35 minutes, or until golden brown. Let stand 5 minutes before serving.

Scalloped Potatoes

2 potatoes per person
1 lg. onion, diced
1 stick margarine

American cheese
1 can Carnation milk
Salt & pepper, to taste

In a glass baking pan, put margarine in bottom, then a layer of potatoes, layer of onion and a layer of cheese. Put butter pats on top. Continue layering until pan is full. Pour on canned milk until level with potatoes. Bake at 350° until done, 45 to 50 minutes.

Capletti

FILLING:

1/4 lb. pork sausage
2 c. grated Parmesan cheese
2 c. grated bread

1 egg
1/2 tsp. salt & pepper
7 T. chicken broth

Mix and knead.

DOUGH:

3 c. flour
3 eggs

7 T. water

Mix and knead. Mix and roll thin. Cut into strips. Cut into squares and seal. Boil in chicken broth. Yield: 10 servings.

Egg Drop Soup

6 eggs
Flour

Salt, to taste
Pepper, to taste

Beat eggs. Mix enough flour to make a thick pancake batter. Drop spoonfuls in boiling chicken broth. Boil until done.

Egg Noodles

12 eggs
Up to 5 lb. flour

Salt, to taste

Beat 12 eggs; add flour gradually (I like to add a cup at a time), until dough is thick and sticky. Then pour out onto floured table. Knead dough, adding flour a bit at a time. Dough should be soft and firm, but not sticky. Cut dough into fourths. Take a section and roll it out. The thinner the dough, the quicker it cooks. I like to roll it out to about 1/8-inch thick. Then flour the top of the dough if it is sticky at all. Roll it up to make a long sausage. Now take your knife and slice them about 1/4- to 1/2-inch thick. Unroll the noodles as you cut them to dry. Store in refrigerator. Add to boiling water and cook until tender. Drain and add butter, salt and pepper, to taste.

Ravioli

NOODLES:
12 eggs
1/2 tsp. salt

Flour

FILLING:
2 lb. breakfast sausage
1 lb. grated Parmesan cheese
3 eggs
1 c. bread crumbs

Salt & pepper, to taste
1/8 tsp. ground nutmeg
1 tsp. parsley

Noodles: In the center of your table, pour 2 1/2 pounds of flour out. Make a small hole in the middle. Crack 12 eggs in a small bowl as to check for any shells. Pour into the center of the hole in the flour. Start bringing the flour in a little at a time as to mix it. Keep adding flour until all flour is incorporated. Add up to 2 1/2 pounds more. Mix and knead until firm dough is achieved. Set aside, covered, until filling is made.

Filling: Brown sausage and drain off grease. Cool and add rest of ingredients; mix well. Roll out dough and cut into 3-inch squares. Put a small amount of filling into it. Fold over and seal. Sometimes you have to put a little water on the edges to get a good seal. Now take a fork and press in the edges all the way around. Keep going until all the dough is done. Set the ravioli aside to dry. I like to leave them overnight. Wrap them in a feedsack towel and put in the refrigerator if you are going to use them that day. If not, put them in a freezer bowl, seal and freeze them.

They are great to have for a quick meal.

Ravioli Sauce

2 sticks butter
3 lg. cans tomato paste
2 lg. cans water

1 1/2 c. sugar
Salt & pepper, to taste

Melt butter and add paste, water, sugar and salt and pepper, to taste. Let warm up slowly as to not scorch. Let simmer for about 20 minutes.

This is a very rich sauce.

Sweet Potato-Cheese Ravioli

2 sm. cans sweet potatoes
2 c. Parmesan cheese

Salt & pepper, to taste
1 c. bread crumbs

Drain sweet potatoes. Mash them and add grated or powdered cheese. Mix in bread crumbs. If mixture is dry, you can add an egg. Salt and pepper, to taste. Follow the recipe for the ravioli dough and same process. These are also great boiled and fried in butter.

Freezer Bread and Butter Pickles

8 c. sliced cucumbers
2 1/4 c. sugar
1 lg. onion
1 1/4 tsp. celery seed

1 1/4 tsp. mustard seed
1 c. vinegar
1/2 tsp. salt

Peel cucumbers. Slice thinly and set aside. Peel onion and slice very thin. Add remaining ingredients. Refrigerate, stirring 2 times a day, for 4 days. At the beginning of the fifth day, put into freezer containers. Pour in the remaining juice to cover the cucumbers and onions; freeze. Take out as you need them.

Real Baked Beans

1 pt. pinto beans
1/2 gal. water
1 1/2 lb. bacon

3 T. molasses
1 c. brown sugar
Salt & pepper, to taste

Boil the beans and the water until shells split off and are tender. Drain beans and wash them off to get the skins off. Continue this until clean.

In bean crock, put beans back in and cover with water. Add browned bacon, molasses, sugar and salt and pepper, to taste. Bake at 350° until bubbling. Add boiling water to keep from getting too dry.

Cakes, Cookies, Pies, Pastries & Candies

Cream Rolls

9 eggs
1 3/4 c. flour
1 1/2 c. sugar
2 tsp. baking powder

1/4 tsp. salt
1 tsp. vanilla
1/2 tsp. almond extract

FILLING:
1 c. milk
3 to 4 T. flour
1/2 c. butter

1/2 c. butter-flavored Crisco
1 1/2 c. sugar
1 tsp. vanilla

Separate eggs and set yolks aside. Beat egg white until soft, stiff peaks and add 1 cup sugar gradually. Beat until stiff peaks are achieved.

In a separate bowl, beat yolks, vanilla and almond flavoring. Add 3/4 cup sugar. Combine the 2 egg mixtures gently. Sift flour, salt and baking powder in bowl. Combine the egg mixture gently to the dry ingredients. Place parchment paper on 2 cookie sheets. Divide dough into half and place onto the parchment-lined cookie sheets. Spread out evenly. Bake at 375° until done.

After baked, spread powdered sugar onto feed sack towels and put baked product onto towel; roll it up.

Filling: Cream butter, Crisco and sugar. In pot, cook flour and milk. Now add the sugar-butter mixture to cooked milk mixture. Beat well. Cool mixture. Unroll baked layers and spread with filling; Reroll layer. Slice cream roll about 1-inch thick and serve.

These freeze nicely.

Baked Peaches

6 to 8 peaches
1 c. sugar

1 T. lemon juice
Cinnamon

Peel peaches and place in 9x13-inch pan. Fill halfway with water. Sprinkle with sugar and cinnamon. Add a bit of lemon juice to the water and bake, covered, until tender.

Pluck 'ems

3 loaves bread dough, thawed
2 c. brown sugar
1 sm. pkg. vanilla or butterscotch
   pudding

1 stick butter
1 1/2 tsp. cinnamon
Chopped nuts

Mix brown sugar, cinnamon and pudding in bowl and set aside. With a pair of scissors, cut off the pieces of dough about the size of a walnuts, and dip them into the melted butter. Roll in dry ingredients. Place on greased bundt pan, or a round cake pans works, too. Sprinkle with nuts as you fill the pan. Do not overfill, as with store-bought dough, they really rise and will roll out into your oven. Bake at 350° until done.

Gingerbread

1/2 pt. molasses
1/2 lb. brown sugar
1/2 lb. butter
7 eggs

1 tsp. ginger
1/2 tsp. cinnamon
1 pkg. yeast
Milk

Cream butter, sugar and molasses. Add eggs, one at a time. Mix in the spices. Now take 1/4 cup milk and warm it to lukewarm, and dissolve yeast. Add to batter. Now add enough milk to make a thick batter. Pour into cake pan and bake at 350° until done.

Buckwheat Pancakes

1 pkg. yeast
2 c. warm water
1 c. scalded milk
3 T. brown sugar

2 c. buckwheat flour
1 c. white flour
1/2 tsp. salt

Dissolve yeast and sugar in warm water. Add remaining ingredients and beat until smooth. Cover and set aside, about 1 1/2 to 2 hours. When it has risen, mix again and put on hot griddle.

Cream Pie

3 egg whites
1 c. sugar
1 1/2 tsp. vanilla

1/2 tsp. nutmeg
1/2 tsp. cinnamon
1 pt. heavy cream

Beat egg white until stiff. Beat cream slightly. Add other ingredients. Incorporate egg whites into other mixture. Pour into prepared pie crust and bake at 350° until firm.

Spice Cake

1 1/2 c. buttermilk
1 1/2 c. brown sugar
1/4 c. butter
1 tsp. baking soda
1 tsp. cinnamon

1 tsp. cloves
1 tsp. allspice
1/2 tsp. ginger
3 c. flour

Cream butter and sugar together; add rest of ingredients in the order they are listed. Beat well so that cake will be fluffy. Pour into cake pan and bake at 350° until done.

Almond Cake

8 oz. blanched almonds
1 tsp. vanilla
1 tsp. almond flavoring
8 eggs, separated

10 T. flour
1/4 lb. butter
8 T. butter

Grind almonds to a powder; set aside. Cream butter and sugar; add egg yolks and beat 2 minutes. Add rest of ingredients, except egg whites. Beat whites until stiff peaks and mix both batter and whites together. Put into greased and floured cake pan; bake at 350° until done. Check with toothpick in center of cake. If it comes out clean, it is done.

Dried Apple Cake

1 c. dried apples
1 c. molasses
1/2 c. butter
1 c. sugar
1 egg

1 1/2 tsp. cinnamon
1/2 tsp. nutmeg
1/4 tsp. salt
2 c. flour

Place dried apples in bowl and cover with water. Add 1/4 teaspoon lemon juice. Put in refrigerator overnight. Pour liquid and apples into saucepan. Add molasses and simmer for 2 hours, being careful not to let scorch. Cream butter, egg and sugar. Add cinnamon, nutmeg and salt. Stir in flour and apple mixture. Bake at 350° until done.

Brown Betty

Bread slices
Butter
Blackberries

1 tsp. cinnamon
1 tsp. allspice

Place slices of bread on the bottom of a 9x13-inch pan. Cover with a layer of blackberries. Sprinkle with a little of the spices and 2 tablespoons sugar. Layer another layer of bread, then berries, spice and sugar. Continue until pan is full. Bake, covered, at 350° until bread is browned.
Apples can be used instead of blackberries. Serve with whipped cream.

Kentucky Pudding

1 1/4 c. molasses
1 c. real milk
1 c. raisins

1 c. chopped dates
2 1/2 c. flour
2 tsp. baking powder

Mix all ingredients and mix well. Steam pudding for about 2 hours, or until set.

Bread Pudding

2 c. bread crumbs
2 c. sugar
3/4 c. butter

1 qt. milk
7 eggs
1 c. favorite jam

Mix the first 5 ingredients, except the egg whites. Pour into baking dish and bake at 350° until set and firm. Remove from oven and spoon jam on top of pudding. Beat egg whites until stiff meringue and put on top of jam; bake again until peaks brown.

Custard

1 pt. real milk
4 eggs
1 1/2 tsp. vanilla

2 tsp. almond extract
1 c. sugar

Beat sugar and eggs until light. Scald milk and pour over egg-sugar mixture. Put back on stove and cook until very thick, stirring often. Be careful not to scorch. Remove from the heat and add the vanilla and almond flavoring. Let cool and refrigerate.

Coconut Pudding Pie

5 egg whites
1 1/4 c. sugar

1/2 c. butter
Graham cracker crust

Beat egg whites until stiff; gradually add sugar. Melt butter and let cool slightly. Incorporate into the egg whites. Add the milk of 1 coconut and mix well. Pour into graham cracker pie shell and top with coconut. Chill 2 hours.

Cherry Cobbler

1 white cake mix

2 cans pie filling

Mix cake per directions on the box. Pour into greased and floured 9x13-inch pan. Spoon pie filling onto and into the prepared cake mix in the pan. Bake at 350° until cake that swells up around the filling turns golden brown. Serve warm with ice cream or Cool Whip.

Cream Branches

2 pkg. yeast
1 c. warm water
1 c. scalded milk
1/2 c. margarine
2/3 c. sugar

2 eggs
1/2 tsp. salt
1 tsp. vanilla
1 tsp. almond flavoring
6 1/2 c. flour

FILLING:
3 T. flour
1 c. milk

1 c. sugar
1 c. butter

FROSTING:
3/4 c. brown sugar
5 T. butter
2 T. milk

1 tsp. vanilla
2 1/2 c. powdered sugar

Dissolve yeast in warm water. Let scalded milk cool slightly. Add other ingredients to the scalded milk and incorporate yeast mixture. Let rise and punch down. Roll out to 1/2-inch thick and cut into bars; let rise again. Bake on top rack at 375° for 7 to 9 minutes.

Filling: Cook flour and milk; beat in sugar and butter. Slit bars on one end and fill with pastry bag.

Frosting: Cook brown sugar, butter and milk and cool. Add vanilla and powdered sugar. Frost bars.

Shoo-Fly Pie

CRUMB MIXTURE:

2 c. flour
3/4 c. brown sugar
2/3 c. butter

1/2 tsp. nutmeg
1 tsp. cinnamon

SYRUP MIXTURE:

1 1/2 c. molasses
1 c. brown sugar
3 eggs, beaten

1 1/2 c. hot water
1 1/2 tsp. baking soda, dissolved in
    hot water

Mix crumb mixture until crumbly and all incorporated. In separate bowl, mix syrup mixture ingredients together and beat for 1 minute. Pour mixture into 2 unbaked pie shells. Top with crumb mixture and bake at 350° until firm and set.

Almost Pecan Pie

3 eggs, beaten
3/4 c. dark corn syrup
1/4 c. melted butter
1/4 tsp. salt

1 c. quick oats
3/4 c. sugar
1/2 to 1 c. nuts

Beat eggs and sugar together. Add corn syrup and butter; mix well. Sprinkle in the salt; mix well. Gradually add the oats and nuts; beat 1 minute. Pour in a greased deep-dish pie plate. Let rest for 10 minutes before you put on a cookie sheet and bake at 350° for 45 to 50 minutes.

Peanut Butter Pie

CRUMBS:

1/2 c. peanut butter

3/4 c. powdered sugar

PUDDING:

1/2 c. sugar
2 c. milk
1 T. flour
3 egg yolks
2 T. cornstarch

1 T. butter
1/2 tsp. salt
1 T. vanilla
2 T. peanut butter

You will need a 9-inch pie pan.

Stir together the sugar, flour and cornstarch and salt with enough milk to make a paste. Heat the rest of milk until hot, not skinned. Add small amount to beaten egg yolks. Then incorporate rest of hot milk. Boil 1 minute, or until thickened, and add butter and vanilla. Press crumb mixture into bottom of 9-inch pie plate. Pour pudding mixture over crumb crust. Refrigerate 1 hour before serving.

Caramel Nut Pie

1 c. sugar
1 heaping tsp. flour
1 c. milk
1 c. heavy cream
2 eggs

1/2 tsp. vanilla
1/2 tsp. maple nut flavoring
Pinch of salt
1/4 c. chopped nuts
1 tsp. almond extract

Mix flour, sugar, egg yolks, flavoring and salt. Add enough milk to make a smooth batter. Bring milk to a boiling point. Add to the other ingredients. Beat egg white to stiff peaks. Fold into mixture. Pour into unbaked pie shell. Sprinkle chopped nuts on top. Bake at 375° for 10 minutes. Then turn down temperature to 325° until pie is almost set.

For elderberry pie, omit maple nut flavoring and add 1 teaspoon vanilla. Press a single layer of elderberries on top of pie instead of nuts.

Buttermilk Brownies

1 stick butter
1 c. water
1/2 c. cocoa
1/2 c. Crisco

1/2 c. buttermilk
2 eggs, beaten
1 tsp. baking soda
1 1/2 tsp. vanilla

FROSTING:
1 stick butter
1/2 c. cocoa
1/3 c. buttermilk
1 lb. powdered sugar

1 c. nuts
1 tsp. vanilla
Pinch of salt
1 tsp. almond extract

Boil the first 4 ingredients. Pour over the next 3 ingredients. Add buttermilk, eggs, baking soda and vanilla. Bake at 400° for 20 minutes.

Frosting: Boil and cool butter, cocoa and buttermilk. Add powdered sugar, nuts, vanilla and salt. Mix well and frost brownies.

Cherry Coffeecake

BOTTOM LAYER:

1/4 lb. margarine
1 c. sugar
2 c. flour

2 tsp. baking powder
Egg
Milk

TOP CRUMBLE:

1 c. flour
1 c. sugar

1/4 lb. butter

Put egg in measuring cup and add enough milk to make 1 cup of liquid. Mix all ingredients and put into the bottom of a 9x13-inch pan. Cover with a layer of cherry pie filling, to within 1/2-inch from edges. Mix ingredients for top and crumble on the pie filling. Bake at 350° for 30 to 40 minutes.

Almond Cream Cheese Pound Cake

1 yellow cake mix
1 pkg. cream cheese
4 lg. eggs
1/2 c. water

1/2 c. sugar
1/2 c. oil
1 tsp. vanilla extract
1 1/2 tsp. almond extract

Spray 10-inch tube pan and flour dust it. In a large bowl, mix cake mix, cream cheese, sugar, oil, vanilla and almond. Beat for 1 minute. Turn on high and beat 2 more minutes, scraping side as needed. Pour batter into prepared cake pan. Bake for 35 to 40 minutes at 350°. Let cool on rack for 20 minutes.

Bread Pudding 3

1 hotel pan of cubed dry French
   bread

1 c. apricots
1 c. raisins

CUSTARD:

2 qt. heavy cream
1 lb. sugar
2 1/2 T. cinnamon

1 T. nutmeg
1 tsp. allspice
30 eggs

Pour custard mixture over the cubed bread and fruit. Top with 1/2 pound thinly-sliced butter and bake at 350° in a hot water bath for 2 hours.

Recipe can be condensed.

Flan

CARAMEL TOPPING:

1/3 c. sugar

2 T. water

CUSTARD:

1 1/2 tsp. vanilla extract

1 c. milk

1 c. cream

1/3 c. sugar

2 lg. eggs plus 2 yolks

Place oven rack on lowest position. Preheat oven to 325°.

In a medium saucepan, mix water and sugar and boil until swirling and sugar is dissolved. Boil 5 to 7 minutes. Color of syrup should be amber. Pour into 4 cups, 6 to 7-inch diameter souffle dishes. Make sure to cover the bottom of the dish.

To make the custard, heat milk and dissolve sugar. Add vanilla. Beat egg yolks and add to vanilla mixture. Be sure to add a bit of the vanilla mixture to the egg yolks, a little at a time, to temper the eggs. Then add back to the vanilla mixture. Mix carefully to avoid bubbles in the custard. Divide the custard into the 4 cups and pour over the syrup. Place the dishes in a glass baking dish and add enough boiling water to cover halfway up the dish. Bake custard 50 to 60 minutes, or until a knife is inserted and comes out clean. Remove the dish from the pan of water and cool custard. Chill before serving and will last refrigerated up to 2 days.

Pumpkin Pie

3 pie pumpkins

2 tsp. pumpkin pie spice

1/4 tsp. salt

1 can Pet milk

1 pkg. vanilla pudding

3 eggs

1 c. brown sugar

1 tsp. vanilla

1 tsp. cinnamon

Cut pumpkin into fourths and boil until tender. Spoon out and add enough canned milk until creamy, but not soupy. Beat eggs and add to the mixture. Add the rest of the ingredients; hand mix well. Pour into prepared pie shell and bake at 350° until done.

Angel Cake

1 c. sifted flour
3/4 c. sugar
12 egg whites
1/4 tsp. salt

1 tsp. vanilla
3/4 c. sugar
1 1/4 tsp. cream of tartar

Sift 3/4 cup of sugar and flour 3 times and set aside.

Beat egg whites with cream of tartar, salt and vanilla until it forms soft peaks. Add remaining sugar gradually and keep beating. Gradually sift the dry ingredients into the whites slowly and continue beating until firm peaks are formed. Bake in an ungreased 10-inch tube pan at 375° for 35 to 40 minutes. Invert pan to cool.

Chocolate Angel Cake

Prepare angel cake, substituting 3/4 cup sifted cake flour and 1/4 cup sifted cocoa. Sift cocoa, sugar and flour 5 times.

Chocolate Caramel Toffee Cake

1 German chocolate cake mix
1 (14 oz.) can sweetened condensed
  milk

1 sm. jar ice cream caramel topping
3 c. Cool Whip, thawed
16 oz. toffee chips

Mix and bake the cake per package instruction and bake in a 9x13-inch pan. Poke holes every inch with the handle of a wooden spoon. Drizzle milk over the cake and let stand until it absorbs. Drizzle the caramel on top of that and let stand until it is absorbed. Cover and chill overnight. Serve with Cool Whip and sprinkle with toffee chips.

Chocolate Buttermilk Pie

1 pie crust
1 lg. bag chocolate chips
7 eggs
1 c. buttermilk

1/4 tsp. salt
1 tsp. vanilla
2 1/2 c. sugar

Melt chocolate in microwave at 30-second intervals, stirring each time. Be careful not to burn because chocolate melts at a very low temperature. In a mixing bowl, beat eggs, sugar, buttermilk, salt, vanilla and sugar. Slowly, while still beating, add chocolate. Keep beating for about 3 minutes. Pour into pie crust and let stand for about 10 minutes before you bake it. Be careful to leave room for the pie to rise. Bake about 1 hour at 325°.

Pecan Pie

3 eggs
1 c. sugar
1 c. Karo syrup

1 tsp. vanilla
1 1/4 c. pecans

Beat eggs; add remaining ingredients. Put pie crust into deep-dish pie plate. Put pie crust in freezer. Pour pecan mixture into frozen pie crust. Place pie plate onto cookie sheet. Bake at 350° for 50 to 55 minutes.

Quick and Easy Coconut Cream Pie

2 c. cold milk
2 pkg. coconut cream pudding
1 c. flaked coconut

2 c. Cool Whip, thawed
1 graham cracker crust
1/2 c. toasted coconut

Beat milk, dry pudding and coconut for 2 minutes. Gently beat in 1 cup of Cool Whip. Pour into crust and chill 2 hours. Spread the other cup of Cool Whip on top of the pie. Sprinkle with toasted coconut, if desired.
Makes a pretty presentation.

Buttermilk Pie

5 eggs
2 c. sugar
1 c. flour

1 tsp. vanilla
1 tsp. almond flavoring
1/2 c. buttermilk

Beat eggs; add sugar, vanilla, almond and buttermilk. Beat and gradually add flour. Beat for about 5 minutes on high. Pour into pie crust and leave room for the filling to rise. Let stand for about 5 minutes. Place onto cookie sheet and bake at 350° until done.
My farm hand gave me this recipe from his mama and aunt.

Peanut Butter Blossoms

1 3/4 c. flour
1 tsp. baking soda

1/2 tsp. salt

Cream together:
1/2 c. butter
1/2 c. peanut butter

1/4 c. sugar
3/4 c. packed brown sugar

Add:
1 egg

1 tsp. vanilla

Add to dry ingredients and shape into round balls. Roll into sugar and place on ungreased cookie sheet. Bake at 375° for 10 minutes. When done, place Hershey kiss on top before removing from cookie sheet.

Christmas Nut Balls

1/2 c. butter
4 T. sugar
1 egg yolk
1 T. vanilla
1 T. lemon juice

1 1/4 c. flour
1/4 tsp. salt
1 egg white
1/2 c. nutmeats

Put aside 1 egg white, 1/2 cup nutmeats and maraschino cherries.
Cream butter and sugar; add egg yolk and beat. Add lemon juice, flour and salt and refrigerate dough for a bit to make firm. Form into small balls and roll into beaten egg white. Roll in nuts and top with a cherry. Place on greased cookie sheet and bake at 350° for about 35 minutes. Yield: about 1 1/2 dozen.

Potica

DOUGH:
1 c. scalded milk
1/2 lb. butter

2 T. Crisco

Melt above together.

1/2 c. sugar
3 tsp. salt
5 1/2 c. flour

2 cakes yeast
1/4 c. warm water

Dissolve yeast in water.

3 eggs, beaten

FILLING:
2 lb. ground nuts
2 c. sugar
2 T. cocoa
1 tsp. vanilla
2 T. lemon juice

1/2 c. honey
16 oz. sour cream
3 eggs
1 c. milk, mixed with 1/4 lb. melted
  butter

Refrigerate filling overnight. Put sugar in large bowl and add milk and butter and Crisco. Add eggs, then add yeast mix into flour. Let rise until double in size, about 4 hours. Cut dough into 2 parts; dough will be sticky. Roll on floured table, cover with filling and roll. Put in greased pans and let rise until double. Bake at 350° for 50 minutes, or 35 minutes for smaller pans.

Butter Cookies

1 lb. butter
2 eggs
1 c. sugar

1 qt. flour
1/2 tsp. baking powder
1 tsp. almond flavoring

Cream butter; add sugar, eggs and other ingredients. Mix well. Preheat oven to 425°. Put batter in cookie press and press out cookies. Refrigerate if dough is too soft. Bake 10 minutes. Watch closely, as they burn fast.

Rosky Cookies

DOUGH:

1 cake yeast
3/4 c. cream
1 T. sugar

1 1/4 c. butter
3 1/2 c. flour
4 egg yolks

FILLING:

1 c. ground nuts
3/4 c. sugar

2 unbeaten egg whites
1 tsp. vanilla

Dissolve yeast into warm cream. Add sugar and let stand 20 minutes. Cut butter into flour using a pastry blender. Add egg yolks and combine with cream mixture. Roll out 1/4-inch thick and cut into 2-inch squares. Fill with filling and fold diagonally in half. Place on greased cookie sheet. Brush top with unbeaten egg whites and bake at 325° for 35 minutes, or until brown.

Old-Fashioned Sugar Cookies

1/2 c. lard or Crisco
1/2 c. butter
1 c. sugar
1 lg. egg

1 tsp. vanilla
2 c. sifted flour
2 tsp. baking soda
3/4 tsp. salt

Cream lard, butter and sugar; slowly add egg. Sift together dry ingredients and gradually add to creamed mixture. Add vanilla. Mix well and chill at least 2 hours to overnight. Dough will be soft.

Preheat oven to 375°. Roll into 1-inch balls and place on well-greased cookie sheet. Take the bottom of a glass and dip it into the colored sugars; press down onto the 1-inch balls. Flatten a bit. Bake for 8 to 12 minutes, or until done. Watch, as they burn very easily.

Farmerboy Cookies

2 c. butter
1 1/2 c. packed brown sugar
1 1/2 c. white sugar
2 tsp. vanilla
5 eggs
6 c. flour

2 tsp. baking soda
1 tsp. salt
1 (12 oz.) pkg. chocolate chips
1 (8 oz.) pkg. M&M's
2 c. chopped nuts

Cream butter and sugar; add vanilla and eggs. Beat well and set aside.

Sift dry ingredients and add to creamed slowly. When mixed, add chocolate chips and M&M's. Drop by spoonfuls onto greased cookie sheets. Bake at 350° for about 10 minutes.

Cornflake Macaroons

1/2 c. sugar
1/2 c. coconut
1/2 tsp. salt
2 1/4 tsp. melted butter

2 1/2 c. corn flakes
1 tsp. vanilla
1 egg

Beat egg white until stiff. Fold in remaining ingredients. Place spoonfuls onto greased parchment paper. Place 1/2 maraschino cherry on top. Bake at 325° until tips turn brown.

Sorghum Cookies

3/4 c. shortening
1 1/2 c. sugar
1 egg
1/4 c. sorghum
2 1/2 c. flour

2 tsp. baking powder
1 tsp. baking soda
1 1/2 tsp. cinnamon
3/4 tsp. ginger
1/4 tsp. salt

Mix together. Make balls and roll in sugar and cinnamon mixture (2 teaspoon sugar and 1 teaspoon cinnamon). Add more flour if needed to make dough still enough to make balls. Bake at 350° until done.

Whoopie Pies

4 c. flour
2 c. sugar
2 tsp. baking soda
1/2 tsp. salt
1 c. shortening

1 c. cocoa
2 eggs
2 1/2 tsp. vanilla
1 c. sour cream or sour thick milk
1 c. cold water

FILLING:
1 1/2 T. vanilla
2 T. milk
2 c. powdered sugar
3/4 c. butter

1 tsp. almond flavoring
1 egg white, beaten stiff
2 1/2 T. flour

Cream sugar, salt, shortening, vanilla and eggs together. Sift flour, baking soda and cocoa together. Add the first mixture alternately with water and sour milk. Add slightly more flour if milk is not thick. Drop by teaspoons. Bake at 400° until done.

Filling: Beat egg white, sugar and vanilla. Add remaining ingredients and beat well. Add a few drops of peppermint flavoring instead of vanilla, if desired.

Romance Bars

1 c. flour
1/2 c. butter
3 T. sugar
2 eggs, beaten
1 1/4 c. brown sugar

1/2 c. chopped nuts
1/2 c. coconut
1 T. flour
1/4 tsp. salt
1/2 tsp. baking powder

Mix the first 3 ingredients together like a pie crust. Press firmly in a 9x9-inch pie pan. Bake at 400° for about 8 minutes. Combine the rest of the ingredients; spread over crust. Bake at 325° for 32 minutes.

Chipmunk Bars

1 c. boiling water
1 c. chopped dates
1/2 c. Crisco
3/4 c. butter
1 1/2 c. sugar
2 eggs
1 1/2 tsp. vanilla

1 3/4 c. flour
1/2 c. cocoa
1 tsp. baking soda
1/2 tsp. salt
1/2 c. chopped nuts
1 c. chocolate chips

Pour boiling water over dates and set aside.
Cream shortening, butter and sugar; add eggs and vanilla and beat until light in color. Add flour, cocoa, baking soda, salt and date mixture; mix well. Pour into 10x15-inch pan. Sprinkle nuts and chocolate chips over top. Bake at 350° until done.

Ginger Molasses Cookies

8 T. butter
1 c. white sugar
1 egg
1/4 c. molasses
1 1/2 tsp. baking powder
2 c. flour

1/2 tsp. ground cloves
1 tsp. ginger
1/2 tsp. salt
2 T. cinnamon
1/2 c. white sugar

Cream butter, 1 cup sugar, egg and molasses together. Add all other ingredients, except 1/2 cup sugar, and mix well. Chill and form into walnut-sized balls. Roll into sugar. Arrange on greased cookie sheet. Bake at 350° for 8 to 10 minutes.

Fold-Up Cookies

3 c. + 3 T. flour
2 T. sugar
1/4 tsp. salt
1 c. Crisco
1/2 c. milk, scalded & cooled

1 pkg. yeast
2 tsp. vanilla
Any flavor Solo pastry filling
1 tsp. almond extract

Sift the first 3 ingredients together; cut in the shortening and mix until it resembles cornmeal. Scald milk and cool to lukewarm; add yeast and dissolve. Add egg and vanilla to flour mixture and mix well. Divide dough into 4 parts and roll out on powdered sugar. Cut into 2 1/2-inch squares. Place a heaping teaspoon filling and fold in opposite corners. Place on greased cooking sheet. Let stand 10 minutes before baking. Bake in 350° oven for 10 to 12 minutes. Remove from oven and let cool on wire rack.

Christmas Nut Balls

1/2 c. butter
4 T. sugar
1 egg, separated
1 tsp. vanilla
1 T. lemon juice

1 1/4 c. flour
1/4 tsp. salt
Maraschino cherries
Ground walnuts

Cream butter and sugar; add the egg and lemon juice. Beat well, adding the flour and salt. Dough will be slightly sticky, so refrigerate overnight, or at least 4 hours. Roll dough into small balls.

In a small bowl, beat egg white slightly. Roll the dough ball into the egg whites. Now roll the ball into ground nuts. Place in greased cookie sheet. Press 1/2 a maraschino cherry onto the top of each one. Bake at 350° until done.

Sugar Cookies

3 c. flour
1 tsp. baking powder
1 tsp. baking soda

2 sticks butter
2 eggs
1 1/4 c. sugar

Preheat oven to 375°. Sift dry ingredients together. Cream sugar, eggs and butter. Incorporate flour and butter mixture; beat well. Roll out dough to 1/4-inch thick and cut with cookie cutters. Sprinkle with sugars and place on ungreased cookie sheet and bake 8 to 12 minutes.

Cream Candy

2 lb. candy
1/2 pt. water
1/4 pt. vinegar

1/2 c. butter
1 1/2 tsp. lemon juice

Put all ingredients in a heavy saucepan. Bring to a boil and boil 15 minutes. Pour out onto a buttered wooden bread board. Let it cool slightly and pull to white.

1800's Molasses Candy

1 qt. molasses

1 1/2 tsp. baking soda

Boil the molasses until it comes to a hard, brittle crack when put in cold water. Just before taking it off the fire, stir in the baking soda. Pour onto a buttered wooden cutting board. When almost cooled, pull until white. Cut into pieces.

Maple Ice Cream

6 egg yolks
3 T. vanilla

2 1/2 c. heavy cream
2 c. maple syrup

Beat egg yolks until light. Heat syrup and incorporate into egg mixture. Return to fire and cook until custard is achieved. Pour into bowl and cool, stirring occasionally. Once it is cooled, stir in cream and vanilla. Place in ice cream churn and make your ice cream.

Apricot Candy

2 1/4 c. apricot pulp
2 c. powdered sugar
1 1/2 c. sugar

1 1/2 tsp. cornstarch
3 tsp. lemon juice

Apricot Pulp: Made by stewing dried apricots. Press through a coarse strainer. Add sugar and cornstarch. Mix in the lemon juice and cook until stiff. Cool and pour onto well-greased wood board. Let firm up and cut into squares. Coat with sugar and store.

Cream Candy 2

2 1/2 c. sugar
1/2 c. heavy cream
1/2 c. water

2 T. vanilla extract
1 1/2 c. chopped walnuts
1/4 tsp. cream of tartar

Bring water and sugar to a boil until it becomes clear. Add cream slowly to not stop boiling. Boil until brittle in cold water. Pour out onto buttered wooden cutting board and let cool slightly. Pull until white. Cut into pieces and allow to cure in airtight container for 1 week before eating. It will be really creamy.

White Taffy

1 pt. heavy cream
1 pt. white Karo syrup
4 1/2 c. sugar

1 T. gelatin
1 tsp. paraffin

Pour gelatin in 1/2 cup water to dissolve. Add paraffin about the size of an egg and 1 teaspoon vanilla. Boil to soft ball stage, then add softened gelatin to paraffin. Continue to cook to hard ball stage at 275°. Remove from heat and pour onto buttered 9x13-inch pans to cool.

Buckeye Candy

CANDY:
1 stick softened butter
1 3/4 c. smooth peanut butter

1 1/2 tsp. vanilla
3 1/2 c. powdered sugar

CHOCOLATE COATING:
1 T. vegetable oil

1 (16 oz.) pkg. semi-sweet chocolate chips

Cream butter, peanut butter and vanilla together. Add powdered sugar until proper consistency is reached. Roll candy into 1-inch balls. Place on waxed paper-lined cookie sheet. Chill thoroughly.

Melt chips and vegetable oil together in top of double boiler. Keep chocolate mixture in double boiler over low heat until you dip each candy. Using a toothpick, dip each ball, covering about 3/4 of each candy. Return candy to sheet to cool.

Note: Candy dips best when balls are cold.

Caramel Clusters

1 pkg. Kraft caramels
1/4 c. water

6 c. Rice or Corn Chex cereal
2 c. pecans or walnuts

Microwave caramels and water in microwave-safe bowl at 30-second intervals, until melted and mixed well. Put cereal and nuts in large bowl and pour caramel over top. Mix well and pour on 2 cookie sheets. Bake at 300° for 15 minutes. Stir and bake again for 10 more minutes. Spoon out on waxed paper, cool and break apart.

Noodle Clusters Candy

1 pkg. chocolate chips
1 c. peanuts

3 1/2 c. chow mein noodles

Melt chocolate chips in microwave at 30-second intervals until melted. Pour over peanuts and noodles. Spoon out onto waxed paper.
I also use butterscotch morsels instead of chocolate.

Pie Crust Pastry

1 1/2 c. flour
1/4 tsp. salt

1/2 c. shortening
4 to 5 T. ice cold water

LARGE CRUST OR TOP & BOTTOM:
2 c. flour
2/3 c. shortening

1/2 tsp. salt
5 to 7 T. ice cold water

Sift flour and salt together in a bowl. Cut in shortening. When crumbly, add enough water to hold into a pie dough. The least you work it, the flakier it will be. Roll out and put in pie plate.

Lemon Squares

1 1/4 c. + 4 T. sifted flour
3/4 c. sifted powdered sugar
1/2 lb. room-temp. Butter
4 eggs
6 T. fresh lemon juice

2 c. sugar
1/2 tsp. baking powder
Additional powdered sugar for
   sprinkling on top

Cream butter; add 1 1/4 cups of flour and powdered sugar until a soft dough forms. Spread in 9x13-inch ungreased pan. Press down softly. Bake at 350° for 25 minutes, until light golden brown.

In the meantime, beat eggs, lemon juice and sugar. Mix in the 4 tablespoons flour and baking powder. Mix well. Pour wet ingredients over your crust in the 9x13-inch pan. Bake 25 minutes more. Cool 30 minutes. Sprinkle with powdered sugar and cut into squares.

Dump Cake

Yellow cake mix
1 lg. can cherry pie filling

1 can crushed pineapple, drained
2 sticks butter

Grease a 9x13-inch pan. Dump in the can of cherry pie filling. Pour the crushed, drained pineapple on top. Sprinkle the dry cake mix over the pineapple. Cut the butter into pats and dot them all over the dry cake mix. Put in the oven at 350° until the top is brown.

Notes & Recipes

Breads
& Rolls

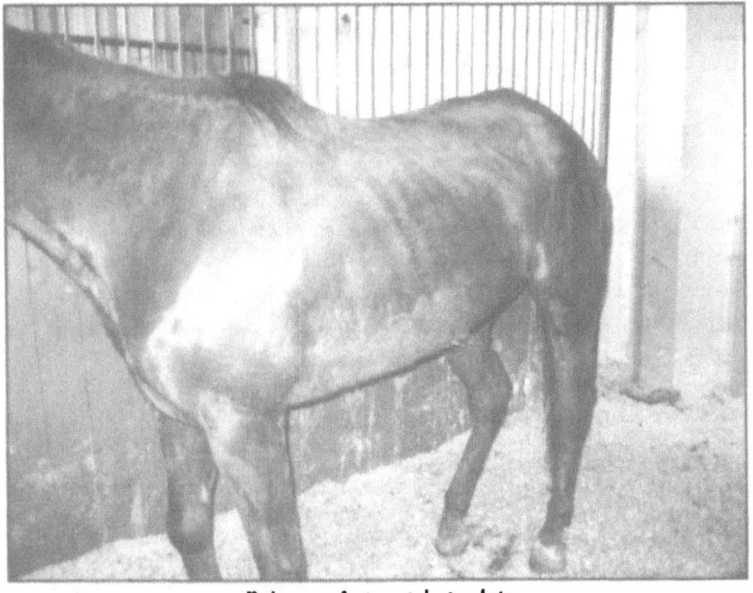

Daisy - unfortunately too late

Daisey

We received a call one summer day that a horse was in trouble. I drove over and the owner's son was there. The owner, after her year-long bout with cancer, had passed away. She had been buried two weeks' prior. He said he was stricken with such sorrow that he just couldn't do it anymore.

My husband and I walked with the man out to his mother's barn. She had employed a man to take care of the animal while she was sick. Well, it was horrible. Daisey was standing in her stall that had so much manure in it that her rear was touching the ceiling of her stall. Every bone she had was screaming out at us. I began to cry. She was so sweet and loving and had given this poor cancer-stricken woman faithful companionship for years. Someone really dropped the ball on this one.

She could barely walk. I didn't think she would make the short trip to the shelter. I called our vet to meet us at the shelter. We had to hoist her up in her stall and give her I.V. fluid. After the fourth liter, she was trying to stand on her own. He put eight inches of shaving in her stall and set her back down on her own feet.

I slept in the barn that night. Giving her a bit of grain every couple of hours. We had her teeth power floated. We soaked alfalfa pellets for her. She chocked on the hay, so we had to feed her leaves of the alfalfa hay and soaked mashes.

After six weeks she was able to go outside. Being so fragile, we put her with another very small skittish mare and foal. We watched them for hours and seemed to be getting along really well. By the end of the day, they were standing side by side waiting to come in for the night.

The spring turned into summer. I noticed that the foal was getting very close to Daisey. I couldn't believe my eyes when I walked down to let them in for the night, but the foal that belonged to the other mare with Daisey was dry-nursing on Daisey. She stood there for it. I was shocked. The three of them were inseparable.

Daisey unfortunately was too far gone. She passed away from our sight four months after she came. We worked so hard to make her well. It was too late for her.

Breads & Rolls

Yeast-Risen Dough Donuts

2 pkg. yeast
1/2 c. warm water
1/2 c. scalded milk
1/3 c. Crisco shortening
1/2 c. sugar

1/2 tsp. salt
3 1/2 to 4 c. sifted flour
2 eggs
Oil, to fry in

Dissolve yeast in warm water and set aside.

In a large bowl, mix scalded milk, Crisco, sugar and salt. When lukewarm, incorporate yeast mixture with milk mixture. Mix well. Add beaten eggs and mix well. Gradually add flour until dough consistency. Let rise and punch down. Roll out and cut. Place in hot oil and fry 2 minutes; flip over. Fry until golden. Glaze or cinnamon sugar or powdered sugar.

Pizza Dough

3 c. flour
1 tsp. salt
1/2 c. vegetable oil

1/2 c. milk
1 tsp. yeast
1/2 c. water

Mix all together and let stand a little before rolling out. Yield: 2 cookie sheets.

Quick Dinner Rolls

2 pkg. dry yeast
1/2 c. warm water
1 1/2 c. warm water
3 T. sugar

1 tsp. salt
2 T. melted butter
4 1/2 to 5 c. plain flour

Mix yeast and 1/2 cup of water and set aside.

In a large, plastic bowl, mix water, sugar, salt and butter; add the yeast mixture. Gradually add the flour, mixing well. Let rise 15 minutes, then work out and let rise 15 minutes again. Bake at 350° until golden brown; brush with melted butter while still warm.

Dinner Rolls

2 pkg. yeast
2 c. lukewarm water
1/2 c. butter
1/2 c. sugar

2 tsp. salt
3 eggs
6 1/2 c. flour

Dissolve yeast in water. Add sugar, salt and 3 cups flour and beat until smooth. Add butter and eggs. Beat well again. Stir in rest of flour and knead well. Place in greased bowl. Cover and let rise until double in volume. Punch down and roll into rolls. Place on greased pan and let rise. Bake at 325° until golden brown.

Mama's Cornbread

1/2 c. sugar
1 egg
2 T. oil
1 tsp. baking soda
1/2 tsp. baking powder

1/3 tsp. salt
1 c. flour
1 c. cornmeal
1 c. buttermilk

Cream the first 3 ingredients; add the rest of the ingredients. Bake in a 9x9-inch pan at 350° for 20 to 25 minutes.

Sour Cream Cornbread

1 c. flour
1 1/2 c. sugar
2 c. cornmeal
1 1/2 tsp. baking soda

1 1/2 tsp. salt
1 1/2 tsp. baking powder
3 eggs
1 1/2 c. sour cream

Mix all ingredients well. Place in muffin papers in muffin pan. Bake at 350° until golden brown.

Sweet Potato Buns

4 sweet potatoes
Enough flour to form dough
1/2 tsp. cinnamon
1/2 tsp. nutmeg

1/2 tsp. ginger
1 tsp. cloves
1 c. sugar
2 pkg. yeast

Peel and boil sweet potatoes until soft. Cool and smash. Mix in enough flour to form dough. Add remaining ingredients and let rise. Punch down, form into rolls and bake until light golden brown. Spread cream cheese frosting on top.

Indian Fry Bread

3 c. flour
1 c. powdered sugar
1/3 c. + 1 T. milk

1 c. cold water
Oil, for frying

Mix flour, sugar, milk and cold water. Mix until a dough is formed to roll out. Roll out and cut with a glass. Place in hot oil and fry; flip over and fry other side. Remove from oil and powder with powdered sugar.

Doughnuts

5 eggs
1 c. cake flour
1/2 c. butter

1 c. milk
4 T. baking powder

Beat eggs until light in color. Cream with the butter. Add flour, milk and baking powder and mix well. You might have to add a bit more flour if dough is sticky. Roll out and cut. Heat up oil and fry doughnuts; flip over and fry other side. Take out and coat with sugar or cinnamon sugar or powdered sugar, and glazes.

Sourdough Biscuits

1 pkg. yeast
1 c. warm water
2 c. buttermilk
3/4 c. oil
1/2 c. sugar

4 T. baking powder
2 tsp. salt
1/2 tsp. baking soda
6 1/2 c. flour

Combine all dry ingredients. Add buttermilk, oil and water. Put into covered bowl. Refrigerate and use as needed. Bake as regular biscuits, at 350° to 375° for 15 minutes.

Angel Biscuits

6 c. flour
3 T. baking powder
1 1/2 tsp. salt
2 T. baking soda
1/2 c. sugar

1 c. shortening
1/4 c. warm water
2 T. yeast
2 c. buttermilk

Mix flour, baking powder, salt, baking soda and sugar together. Cut shortening into flour mixture. Dissolve yeast in warm water. Combine milk and yeast mixture; mix with flour. Roll dough out and cut with biscuit cutter. Dip bottom of cut biscuits in melted butter. Turn biscuit over and bake, butter-side up, at 400° for 12 to 15 minutes.

Lebkuen

1 qt. molasses
1 qt. brown sugar
1 qt. buttermilk
1 lb. lard
2 1/2 T. baking soda, put into
   buttermilk
2 c. raisins

13 c. flour
1 1/2 tsp. cinnamon
1 tsp. nutmeg
1 tsp. ginger
1 T. salt
1 c. chopped dates

Beat lard and sugar together; add molasses. Alternate buttermilk and dry ingredients. Last add nuts, raisins and dates. Let dough stand in cool place for 3 to 4 weeks. Bake like cookies, then store 3 to 4 weeks.
I don't wait that long to eat them.

Hillbilly Bread

4 c. warm water
3 T. yeast
5 c. whole wheat flour
1 T. salt

1 c. warm water
1 c. brown sugar
1 c. oil
10 to 12 c. flour

Mix the first 4 ingredients together and let stand 1/2 hour. Add remaining ingredients and mix well. Let rise, punch down and let rise again. Put into loaf pans and let rise again. Bake at 350° for about 30 minutes, or until brown.

Cinnamon Rolls

1 c. warm water
2 pkg. yeast
1 c. scalded milk
1/2 c. sugar

7 c. flour
3 eggs
1/8 tsp. salt
2 T. cinnamon

CARAMEL NUT ICING:
1 1/2 c. dark brown sugar
1 c. nuts
1/4 c. milk

3/4 c. butter
3 c. powdered sugar

FILLING:
Cinnamon
Brown sugar

Nuts (or see Note)

In a large plastic bowl, add scalded milk and butter. Add sugar. Let cool slightly. Add yeast and eggs, salt and cinnamon. Mix well and incorporate the flour. Let rise and punch down. Divide dough in half and roll out oblong to 1/4-inch thick. Brush with melted butter and sprinkle filling all over to the edges. Roll lengthwise and cut in 3/4-inch-thick slices. Place in greased pans and let rise again to double in size. Bake at 350° until golden brown.

Note: You can customize your cinnamon rolls with nuts, dates, raisins, dried fruit, dried cranberries, etc. let your imagination run.

Caramel Nut Icing: Melt butter and add brown sugar, milk and nuts. Cook over low heat, stirring constantly, until it comes to a boil. Let boil about 1 minute and remove from heat. Let cool and add powdered sugar. Add more or less to achieve proper consistency. Pour over rolls.

Sourdough Starter

2 pkg. yeast
1/2 c. very warm water
2 c. lukewarm water

2 T. sugar
1 T. salt
2 c. flour

Dissolve yeast into 1/2 cup very warm water. In a large bowl, add 2 cups warm water, sugar and salt. Stir in yeast mixture when dissolved. Gradually add flour mixture and mix well. Cover with towel. Leave at room temperature for 2 to 3 days, until begins to ferment. When it becomes foamy and bubbly, it is ready to use. You may use it immediately, or store it in refrigerator. When using the starter, add 1 cup warm water, 1/2 cup flour, 3 teaspoons of sugar for each cup of starter removed. Cover jar and let stand until ready again.

Sourdough Bread

5 to 6 c. flour
2 c. starter
1 c. oil
1 1/4 c. sugar

1 tsp. salt
1 c. milk
4 T. yeast
6 lb. flour

In a large bowl, dissolve yeast and water. Add starter, oil, sugar, salt and milk. Mix well. Stir in 1 pound of flour at a time and beat until smooth. Add rest of the flour slowly and knead well. Let rise and punch down. Put in greased loaf pans and let rise until double in volume. Bake at 350° until light brown.

Whole Wheat Bread

4 c. hot water
1 1/2 c. honey
3/4 c. oil
1 1/2 T. salt

4 T. yeast
4 1/2 c. whole wheat flour
7 to 9 c. white flour

Dissolve yeast in 1 cup water and add sugar. Set aside.
In large bowl, place 3 cups water, honey, oil and salt. Mix well. Gradually add yeast mixture. Mix well and gradually add flour, mixing well after each cup. Knead dough. Let rise and punch down. Place into greased loaf pans. Let rise until double in size. Bake at 350° for 20 minutes, or until golden brown.

Danish Braid

1 c. softened butter
2 c. flour
1/4 tsp. salt
3/4 c. sugar

3 beaten eggs
2 pkg. yeast
1 c. warm water

FILLING:
1/2 c. butter
1/2 c. cream cheese
1 1/4 c. dark brown sugar

1 T. cinnamon
1 1/2 c. nuts
Raisins, dates, dried fruit (opt.)

Filling: Mix the night before and refrigerate, covered.

Incorporate butter, flour and salt; add eggs and sugar. In a small glass bowl, dissolve yeast in warm water,. Add to your dough and mix well. Let dough rise to double in size, then divide into 4 sections. Roll out 1 section to the size of the 9x13-inch pan.

Place filling into the center of the rolled-out dough and spread down center only, using a little over 1/2 cup filling per braid. Now cut strips on the sides and cross over the filling. Bake at 350° for about 20 to 25 minutes, or until light brown.

I like to spread a glaze or powdered sugar over the top. For a real treat, I like to put a maple glaze over the top.

Light Buns

1 qt. real milk
3/4 c. sugar
1/2 c. margarine

1 tsp. salt
2 compressed or pkg. yeast
Flour

Scald milk and add sugar and margarine. Let cool to lukewarm and dissolve yeast. Add flour to make dough. Let rise and pinch down; form into rolls. Place on greased baking dish and bake until light golden brown.

Buttermilk Biscuits

3/4 c. buttermilk
2 c. self-rising flour

Pinch of baking soda
3 T. oil

Mix well until sticky. Roll out and cut. Bake at 350° until light golden brown.

Yeast Dinner Rolls

4 to 4 1/2 c. flour
1/4 c. sugar
2 env. yeast
1 tsp. salt

1 1/2 c. scalded milk
1/3 c. butter
2 lg. eggs

Scald milk and pour into large bowl. Add butter and sugar. Let cool until lukewarm and add yeast and salt. Let cool a bit more and add beaten eggs. Now, add the flour gradually. Let rise. Punch down and make into rolls. Place into greased 9x13-inch pans and let rise again. Bake at 375° for 20 to 40 minutes, until light golden brown.

I like to spray the top of them with butter-flavored spray.

Pancake Syrup

1 c. brown sugar
1 c. white sugar
1 c. water
Pinch of salt
3 T. cornstarch

1/2 c. water
1 c. Karo syrup
Vanilla flavoring
Maple flavoring

Mix the first 4 ingredients and bring to a boil. Mix cornstarch and 1/2 cup water. Stir quickly into boiling sugar mixture. Stir until thickened slightly. Remove from heat and add white syrup and flavorings, to taste.

Pancakes

5 egg whites
2 T. light oil
1 c. milk
1 1/2 c. flour

1/2 tsp. salt
4 T. sugar
2 T. baking powder

Beat egg whites until stiff. In a separate bowl, add remaining ingredients. Fold in egg whites. Fry on hot griddle.

Extremely light and fluffy pancakes.

Gram's Pancakes

2 T. oil
1 egg
1 c. milk
1 c. flour

1/2 tsp. salt
2 T. sugar
2 T. baking powder

Mix oil, egg and milk; add dry ingredients. Fry on hot griddle. Yield: enough for 3 large portions.

These pancakes are very light.

Notes & Recipes

Camp Fire
& Open Fire Recipes

Penelope & daughter

2 baby fox - 1 baby pup - Mom dog took care of them all

For Weak Kids (Baby Goats)

For newborns that won't nurse and/or are weak, try the following...dilute some Karo syrup in two pints of warm water and carefully dribble a little into the kid's mouth. Don't feed too much or too fast. A slow flow nipple for a regular baby bottle will do just fine.

The solution supplies quick, digestible energy, often enough to get the kid strong enough to begin normal nursing. If you don't have any Karo syrup, you might try a little black coffee. It might be enough stimulant to get it going.

If the kid is too weak to nurse, you might try tube feeding. If the kid fails to respond quickly seek medical attention.

If you go out to the barn and the kid is laying out flat but breathing and very weak, I have had good luck to get their body temp back up. I immerse them (keeping their head out) in a bucket or sink of warmer than body temp water. Then I wrap them up like mummies to dry them off and tube feed them.. I mix 2/3 colostrums and 1/3 black coffee. The caffeine gives them a "jolt."

Camp Fire & Open Fire Recipes

Campfire Paper Bag Breakfast

4 slices thick bacon
1 egg

1 paper bag
1 stick

Open lunch bag and place bacon in the bottom, being sure to cover the whole bottom of the bag. Crack your egg and dump out on top of the bacon. Carefully fold the paper bag. Poke a hole in the top of the bag to slip your stick through. Hold stick over hot coals of fire until cooked, being careful to cook, but not burn the bag.

Camp Fire All-In-Ones Breakfast

4 sausage links
3 eggs
1 potato

1/4 c. shredded Cheddar cheese
3 T. milk

Cook sausage and cut into pieces. Slice potato into thin slices. Cook potato in sausage drippings. Drain off grease. Beat the eggs and pour into the potato pan and cook until done. Put sausage back into pan and reheat. Mix milk and cheese and pour over the pan. Cook until cheese is melted and serve with toast.

Camp Fire Foil Breakfast

Hash browns
Eggs
Sausage

Salt & pepper, to taste
Heavy-duty aluminum foil

Take a piece of foil and spray with Pam. Place hash browns and sausage and beaten eggs in foil; season to taste. Fold foil up and make sure well sealed and enough room for expansion. Place in the hot coals of your camp fire. Cook approximately 15 minutes, turning and rotating as needed.

Omelet In a Bag

(Camp Fire Breakfast)

Eggs

Milk

Cheese

Bacon or sausage

Diced onions (opt.)

Sliced mushrooms (opt.)

Chopped green peppers (opt.)

Fry bacon or sausage. Beat eggs, milk and cheese well. Pour into a ziplock bag. Place your meat and goodies in the bag with the eggs. Put a pot of water on to boil. When water is boiling, place ziplock bag into water. Cook until done.

Easy and no clean-up.

Camp Fire Foil Dinner

1/4 lb. burger

1 potato, sliced

1 carrot, sliced

1 onion, sliced

Salt & pepper

Butter

Heavy-duty foil

Spray a piece of foil with Pam. Place your burger patty on the center of the foil. Place potatoes and cooked bacon over patty. Now pick out your favorite vegetables and place on top of the potatoes. Stack up and place a couple of butter pats on top, salt and pepper, to taste. Fold foil up and around your raw dinner, crimping the edges and rolling up, leaving room for expansion. Poke a hole in the top to let out the steam. Place in the hot coals for 20 to 30 minutes. Turn and rotate your pouch as needed.

Campers Pizza

Bread

Butter

Mozzarella cheese

Pizza sauce

Sliced pepperoni

Pie irons

Butter bread on 1 side. Place 1 piece of buttered bread butter-side down in the pie iron. Top with 1 tablespoon sauce, slices of pepperoni and cheese. Now put the buttered bread, butter-side up, and seal the pie iron. Place pie iron in the hot coals, turning and checking often for toasting.

Mmmm Mmmm Good. Kids love this because they can create their own pizza and instead of holding a stick with a hot dog, they're holding on to the pizza!

Campers Pie

Ground beef
Potatoes
Onions
Carrots
Mushrooms

Peppers
Butter
Dry onion soup mix
Heavy-duty aluminum foil

For each pie, spray foil with Pam. Place a thin burger in the center. Slice a small potato thinly on top of the burger. Put 2 butter pats on top and sprinkle with dry onion soup mix. Now create your own vegetable concoction. Wrap up your foil pie securely. Place in the center of the hot coals and cook 30 to 45 minutes, or until done, turning often so as to not burn. Poke a small hole to let the steam escape. When meat is cooked and vegetables are tender, it is done.

Campers Hay Stacks

1 pkg. corn tortilla chips
1 can chilies
1 onion, diced
Lettuce

Tomato, diced
Shredded Cheddar cheese
Salsa

Warm chili in Dutch oven. When hot, put in several bowls. Top with corn chips, onions, lettuce, cheese, tomatoes and salsa. Serve.

Campers Foil-Wrapped Potatoes

Potatoes
Onions
Butter

Salt & pepper, to taste
Heavy-duty foil

Take a large piece of foil and spray with Pam. Place 6 tablespoons butter on foil and top with thin-sliced potatoes, seasoning as you go. Slice onions if desired and place on top of the potatoes. Top with 6 tablespoons butter and salt and pepper, to taste. Fold foil tightly and place in hot coals. Turn often to insure even cooking. Punch hole into top to let steam escape. Cook for 25 to 35 minutes, depending on how many potatoes you sliced. You can check for tenderness, just be careful of the steam. It burns as quick as fire.

Campers Corn On the Cob

Husked corn   Salt
Butter   Ice cubes

Spray heavy-duty foil with Pam. Put 1 ear of corn in the center. Put butter pats on top of the ear of corn and season with salt. Put 2 ice cubes in the pouch and wrap up. Place in hot coals and cook about 15 minutes, turning often.

Campers Crescent Rolls On a Stick

Crescent rolls   Jelly (opt.)
Butter   1 inch green hardwood

Clean off your stick. Wrap the roll in a spiral fashion around the stick. Leave space between the spiral for the heat to reach all the dough. Press the ends to secure onto stick. Hold over the hot coals 15 to 20 minutes. When it is golden brown, slip the crescent off the stick. Top with butter or jelly.

Apple Crisp Campers Style

Apples   1 1/4 c. flour
3/4 c. butter   1/2 c. quick oats
3/4 c. dark brown sugar   1 T. cinnamon

Spray Dutch oven with Pam. Place sliced apples in the bottom of a Dutch oven to fill halfway up. Combine brown sugar, flour, oats and butter. Crumble on top of the apples. Cover and cook about 45 minutes, or until apples are tender.

Peach Cobbler Campers Style

1 lg. sliced peaches   Bisquick
1/4 c. of the heavy syrup

Put the peaches back in the can. Mix the syrup with the Bisquick until it is the consistency of thick batter. Pour over the peaches in the can. Set can in hot coals and cook until hard dumplings are formed.

Campers Pecan Pie

2 c. chopped pecans
3/4 c. packed brown sugar
3/4 c. milk
3/4 c. corn syrup

1/2 c. Bisquick
1/4 c. margarine
4 eggs
1 1/2 tsp. vanilla

Grease your pie plate. Sprinkle the plate with pecans. Beat remaining ingredients until smooth. Pour into pie pan on top of the pecans. Wad up tin foil into 1-inch balls and place down in the bottom of a Dutch oven. Place your filled pie plate on top of the foil balls. Put the Dutch oven into the hot coals, being careful to make sure it is even so it does not tip over. Bake for 50 to 60 minutes. Stick knife in the center of the pie; if it comes out clean, it is done.

Campers Gingerbread

1 (14 oz.) can applesauce

1 gingerbread cake mix

Spray Pam in the bottom of the Dutch oven. Pour applesauce in the bottom of your Dutch oven or cast iron skillet. Sprinkle the dry gingerbread cake mix over the top of the applesauce. Place in the hot coals and cover. Cook until moisture from the applesauce has steamed the cake mix. Let cool approximately 20 to 30 minutes.

Campers Angel Cake

1 whole loaf white bread
1 can sweetened condensed milk

1 pkg. shredded coconut

Slice the bread into 2-inch cubes. Place on stick and dip in the sweetened condensed milk. Roll in the shredded coconut. Toast over an open fire and enjoy!

Campers Coffee Can Ice Cream

1 (1 lb.) coffee can
1 (3 lb.) coffee can
Duct tape

Crushed ice
Rock salt

ICE CREAM:
1 c. milk
1 c. heavy whipping cream

1 tsp. vanilla
Nuts, fruit, chocolate chips (opt.)

Combine ingredients in the 1-pound coffee can. Tape the lid with the duct tape. Place in the 3-pound coffee can. Put crushed ice and 1 cup rock salt around the 1-pound can. Put the lid on the 3-pound coffee can and secure with duct tape. Roll back and forth on the sidewalk or picnic table. Roll for about 10 minutes. Open the 3-pound can and carefully remove the 1-pound can. Open that can and scrape the sides of the can, mixing the ice cream. Reseal. Drain the water off the 3-pound can and replace the 1-pound in the 3-pound can. Put more crushed ice and 1 more cup of rock salt. Seal well with the duct tape. Roll back and forth for 5 to 10 minutes, or more, depending in how firm you want your ice cream.

Camp Fire Cream Puffs

1 can Pillsbury Grands biscuits
Lg. marshmallows

1 stick
1 camp fire

Open up biscuit and stuff marshmallow inside. Close the biscuit up so you can't see the marshmallow. Put it on your stick and roast until golden brown, or until desired.
Enjoy!!

Camp Fire Pineapple Surprise

Plain cake donuts
Pineapple rings
Brown sugar

Butter
Pam spray

Spray foil with Pam. Slice a donut and place 1/2 on foil. Put a pineapple ring on top of the donut half. Put brown sugar around the pineapple ring. Cut a butter pat and place on top of the brown sugar. Put the top of the donut on. Wrap the foil up and put on a rack over the camp fire. Cook about 5 minutes, checking often, making sure it does not burn.

Sauces, Soups, Salads, Puddings, Beverages & Jams

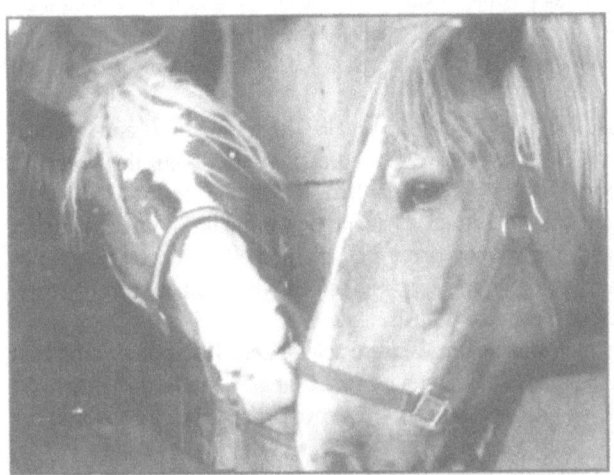

Howdy neighbor

Cheyenne's lazy-grazy days of summer

They Call Me Vida

The call: Hello, I am moving and I can't take my horse and pony. Can you help? Sure I can. She gave me directions. I loaded the trailer and the farmhand and off we went.

It was a beautiful huge house. I saw the woman come out and told me to drive around to the back and go out to the barn.

Everything seemed ok. I walked up to the barn and I could see the two horses in a very small pasture. Vida had a blanket on and that seemed a bit odd for it was April. I asked her why and she said it was cold the night before and she hadn't removed it yet. But she said just leave it on for it goes with him.

He, Vida, was an old retired racehorse. His knees were huge and he could hardly bend them. His hooves were so long that we had to cut them off with a hacksaw before the farrier could trim him. It was hard to load him for he couldn't step up. I backed it up to a hill and he walked down into the trailer.

We then turned our attentions on the little pony. Fat and sassy and not knowing what was going on she appeared to be toying with us in the pasture. Running and ducking our every move.

Finally the old motivator grain. We were able to get a halter on her and load her next to the old man, Vida. She gave us all the hay and feed that she had and grain. All horse gear and saddles. We drove off and she had a tear in her eye.

Getting back, we removed the old man's blanket. I see why he was covered, he was a skeleton! I know that he is old but there are things to do that can help keep weight on. I called our vet and she came out. Blood was drawn, manure taken, floated his teeth, which were worn to nothing.

The farrier came in and tried to trim his hooves, but he couldn't bend his front legs, his knees were too bad. He tried but the old man just couldn't. He was only able to do the back ones. You could see the old man, Vida, really tried.

With soaked alfalfa pellets and senior feed, he started to put on weight. It was the best for the old man to just walk around our shelter. When we tried to stall him, he would go down and was not able to get up. There were many times that we thought this was his time. But he kept on going. Our vet talked to us about putting him down before the winter to save ourselves the trouble of putting him to rest in the frozen conditions. I said if he isn't in any pain, he earned the right to tell us when it was time.

It was November 3 and he laid down for the last time. We sat with him on the ground as he took his last breath. He looked up in our eyes as if to say thank you.

Sauces, Soups, Salads, Puddings, Beverages & Jams

Easy Strawberry Jam

6 c. sugar
1 1/2 qt. ripe strawberries

1 1/2 tsp. alum

Pour sugar and berries in large pot and bring to a boil. Boil about 10 minutes, being sure to stir constantly. Take off heat and add alum. Pour into hot jars and seal.

Pear Butter

25 ripe pears
1 1/4 c. water
5 c. sugar

1 tsp. orange rind
2/3 c. orange juice
1 tsp. ground nutmeg

Boil pears over medium heat until soft, 30 to 40 minutes. Press through sieve. Take the purée and add remaining ingredients. Cook over medium heat, stirring frequently, until thick, about 1/2 hour. Remove the foam off the top and pour into hot jars. Hot bath them for 10 minutes.

Quick and Easy Apple Butter

13 c. applesauce
7 1/2 c. sugar
3 tsp. cinnamon

2 tsp. ground cloves
2 pkg. strawberry Kool-Aid

Mix all ingredients in large canning pot. Bring to a boil. Be careful to not let it scorch. Simmer boil for about 25 to 35 minutes, or until it thickens. Pour contents into canning jars and seal.

Chili Soup

1 lb. hamburger
1/2 c. chopped onion
1 lb. kidney beans
1 qt. tomato juice

1/2 c. brown sugar
1/2 c. ketchup
1/2 tsp. chili powder, or to taste
Salt & pepper, to taste

Soak and cook beans until tender. Fry hamburger and onions together until brown. Add tomato juice, ketchup, sugar and seasonings. Put together and simmer for 30 minutes. Thicken slightly with flour or cornstarch.

To Can Chicken Soup

4 chickens, cooked & deboned
1 gal. cooked noodles
1/2 gal. celery, chopped & cooked
  separately

1/2 gal. carrots, chopped & cooked
  separately
1/2 gal. chopped potatoes, cooked
  separately

Mix in salt and pepper, to taste. Put into jars and cold pack for 2 hours.

Chicken Noodle Soup

1 gal. chicken broth
1 qt. cooked, deboned chicken
1 pt. celery
1 lg. onion, chopped

1 pt. corn
1 pt. cooked rice
1 qt. raw noodles
Salt & pepper, to taste

Bring broth, celery and onion to a boil. Cook until tender. Add chicken, cooked rice and corn. Salt and pepper, to taste. Add noodles and bring to a boil again. Simmer until noodles are tender. Serve before noodles get mushy.

Note: Water and chicken base can be used for broth, about 1 rounded tablespoon per quart of water.

Turkey Chowder

4 c. chicken broth
Chicken or turkey, cooked &
  deboned
Salt & pepper, to taste
1 med. onion, chopped
1/2 c. carrots

1 c. celery, chopped
1 c. potatoes, chopped
2 c. milk
6 T. flour
4 T. butter
1 c. shredded Cheddar cheese

Cook all vegetables in sauté pan until tender. Add broth, milk, flour and butter. Cook until thickened. Add cheese and meat last.
Use Thanksgiving leftover turkey.

Cheesy Chicken Chowder

4 c. cooked, cubed chicken
4 c. broth
1 c. celery
4 c. potatoes
1 c. carrots, chopped

1 lg. onion, chopped
2 c. cubed Velveeta cheese
4 c. milk
4 heaping T. flour

Cook vegetables in broth until tender. Add cheese. When melted, add milk and bring to a boil. Thicken with flour mixed with some milk. Add chicken last. Heat and serve. Yield: 1 gallon.

Cheeseburger Soup

1 lb. hamburger, browned & drained
2 c. cooked rice
1/2 c. carrots
1/4 c. onions, chopped
1/3 c. celery, chopped

3 c. chicken broth
2 (11 oz.) cans Cheddar cheese soup
2 soup cans of milk
1 (8 oz.) ctn. sour cream
Salt & pepper

I use water and chicken soup base. I make my own cheese soup (below).
Combine broth and vegetables; simmer 10 minutes. Add hamburger, rice, soup and milk. Add sour cream last. Do not boil. If it gets too thick, add water.
This is very good on a chilly fall day, or a cold winter evening.
Cheese Soup: 6 tablespoons butter, 6 tablespoons flour, 1/2 teaspoon salt, 2 cups milk, 1/2 teaspoon dry mustard and 1 cup cheese, either Cheddar or Velveeta. Heat, stirring constantly, until thick.

Cheddar Chowder

2 c. water
2 c. potatoes
1 c. carrots, chopped

WHITE SAUCE:
1/4 c. butter
1/4 c. flour
2 c. milk

1 c. cubed ham or chicken
1/4 c. chopped onion
Salt & pepper, to taste

2 c. grated cheese, either Cheddar or
    Velveeta

Cook the first 7 ingredients until tender. Cook sauce. When thickened, add cheese. Then add to vegetable mixture.

This recipe is very flexible.

Bean and Bacon Soup

1 1/2 lb. navy beans
5 qt. water
20 slices bacon, fried
6 c. chopped onion
4 c. chopped celery
2 c. chopped carrots

4 c. chopped potatoes
8 tsp. salt
2 tsp. pepper
2 bay leaves
5 c. tomato juice

Soak beans overnight; drain water. Add 5 quarts water and boil 5 minutes. Remove from heat and let stand for 1 hour. When hour is up, add fried bacon and bacon grease. Simmer at least 1 hour, then add rest of ingredients. Simmer another hour, or until vegetables are soft. Put through blender. Note to can the soup pressure for 90 minutes at 10 pounds. Yield: 8 quarts.

Fall Soup

1 lb. ground beef
1 c. chopped onion
4 c. water
1 c. chopped carrots
1 c. diced celery
2 c. cubed potatoes

Salt, to taste
1 tsp. bottled brown Kitchen
    Bouquet sauce
Pepper, to taste
1 bay leaf
1/8 tsp. basil

In a soup pot, brown meat. Drain off fat. Cook the onion with the meat and add remaining ingredients. Simmer until tender, about 30 minutes.

Mint Tea

1 c. water

2 1/2 c. sugar

2 c. bruised, chopped mint leaves

Boil water and sugar about 15 minutes. In the meantime, smash and chop up your mint leaves. Pour boiling syrup over mint leaves. Squeeze 2 lemons and add to tea. Grate the rind and add to mixture as well. Let stand overnight.

In the morning, strain and pour over crushed ice or equal parts ginger ale and tea.

Lemon Pop

2 pkg. yeast

2 lb. sugar

2 oz. ginger root, grated fine

8 qt. boiling water

2 1/4 oz. cream of tartar

8 lemons

Add boiling water to sugar and crushed ginger. Add tartar. Let cool to lukewarm. Add yeast to 1 cup lukewarm water and dissolve. Add to other mixture. Cut lemons in half and squeeze out juice into the pot. Grate rind and add to pot. Cover and let stand 10 to 12 hours. Strain and pour into bottles. Store in a cool, dry place. Pour over ice or 1/2 juice to 1/2 ginger ale.

Dandelion Wine

1 gal. boiling water

4 qt. dandelion flowers

5 lb. brown sugar

2 lemons & rinds

2 oranges & rinds

Pour 1 gallon boiling water over flowers. Let stand 24 hours. Strain. In a large pot, boil brown sugar, with the juice of 2 lemons and 2 oranges. Grate the rind of the fruits and put in the pot also. Boil this 10 minutes; strain. Now mix the strained juices and waters that soaked overnight and boil 10 minutes. Dissolve 2 packages yeast and add to your mixture when it has cooled to lukewarm. Pour into bottles and lightly cork. When the processing has finished, put 1 raisin in each bottle. Cork tightly. Store in a cool, dry place.

Fruit Punch Before Refrigerators

7 lemons
4 oranges
1 pineapple

1 qt. strawberries
1 jar maraschino cherries

Squeeze 7 lemons and 2 oranges. Slice the other oranges to put into punch bowl. Slice strawberries and cherries and grate up pineapple and save juice. Mix all together and add 1 pound sugar. Let stand overnight to chill and serve.

Bread Pudding 2

1 pt. milk
1 c. bread crumbs
3 eggs, separated
2 T. melted butter

1/4 tsp. nutmeg
1/4 tsp. cinnamon
1/4 tsp. ginger
1 tsp. baking powder

Mix all ingredients except egg whites. Beat egg whites until stiff. Fold into each other. Pour into cake pans and sprinkle with coconut. Bake at 350° until done.

Potato Pudding

4 med. potatoes
3 eggs, separated
1 1/4 c. sugar
1/4 c. flour

1/2 tsp. salt
1 c. heavy cream
1/2 fresh lemon, juiced & rind

Peel, boil and mash potatoes. Cool.
Mix egg yolks with potatoes and set aside.
Beat egg whites and gradually add sugar; beat to stiff peaks. Add rest of ingredients and pour into baking dish. Bake at 350° until firm.
Best served with fresh berries.

Old-Fashioned Eggnog

1 qt. heavy cream
4 egg yolks

6 T. sugar
1 c. rum

Beat egg yolks and sugar until light and smooth. Beat cream until it slightly thickens. Incorporate into the egg sugar mixture and beat again. Pour in the rum and beat 1 minute. Refrigerate.

Tomato Chutney

40 ripe tomatoes
7 onions
7 bell peppers
2 hot peppers

2 1/2 tsp. salt
2 T. black pepper
2 1/4 c. vinegar
8 c. sugar

Peel, core and dice tomatoes. Chop onions and peppers. Put all ingredients in large canning pot and simmer until thickens. Pour into canning jars and seal.

Homemade Ketchup

10 qt. tomato juice
11 c. sugar
3 1/4 c. vinegar
1 onion

1/4 c. salt
2 tsp. catsup spice
1 box clear jell

If you can't find tomato spice, combine 1 teaspoon each:
Cinnamon
Ground cloves
Ground nutmeg

Ground ginger
Allspice
Dry mustard

Cook tomatoes and let cool; drain off liquid. Put into food processor. Process the onions. Mix in all ingredients. Cook 2 hours, being careful to not let the bottom scorch. Thicken with clear jell. Cook 30 minutes, stirring often. Canning jar process.

French Dressing

1 can tomato soup
1 1/2 c. vegetable oil
1 1/4 c. sugar
1 c. vinegar

2 1/4 T. Worcestershire sauce
Salt & pepper, to taste
1 tsp. prepared mustard
1 sm. onion, grated

Place all ingredients into food processor. Beat until thick. Store in refrigerator.

Sweet and Sour Dressing

1 c. sugar
1 c. oil
1 1/2 T. salad dressing
2 tsp. prepared mustard
1 sm. onion, chopped

1/4 c. water
1/4 c. vinegar
Salt & pepper, to taste
1 tsp. celery seed
1 tsp. mustard seed

Mix all ingredients in food processor until thick and creamy. Store in refrigerator.

Pineapple Cherry Salad

1 can crushed pineapple
Water
1 (3 oz.) pkg. lemon Jello
1 c. heavy cream

1/4 c. sugar
1 c. cottage cheese
1/2 c. cut-up maraschino cherries

Drain pineapple and set juice aside. Pour juice and water to make 1 1/3 cups liquid into a saucepan. Bring to a boil. Place Jello into glass bowl and pour boiling water/juice combo over it and dissolve Jello. Cool slightly until it starts to thicken.

In mixing bowl, whip heavy cream. When thickened, gradually add sugar. Fold cooled Jello mixture into whipped cream. Stir in pineapple, cottage cheese and maraschino cherries; mix well. Pour into pretty bowl and chill overnight.

Aunt Josephine's Salad

1 lg. ctn. Cool Whip
1 sm. pkg. lemon Jello
1 sm. pkg. orange Jello
1 lg. ctn. cottage cheese

1 c. mandarin orange slices
1 c. pineapple tidbits, drained
1 c. pecans or walnuts
1 c. sliced bing cherries

Mix all ingredients together and pour into pretty bowl. Chill at least 3 hours.

Easy.

Raspberry-Nut Salad

2 (3 oz.) boxes black raspberry Jello
2 1/4 c. boiling water
25 lg. marshmallows, melted in
  boiling water
1 c. crushed pineapple

1 c. heavy cream
1 1/2 pkg. softened cream cheese
1 c. chopped walnuts
1 c. bing cherries, cut in half

Mix Jello, boiling water and melted marshmallows until thick. Beat cream cheese and cream together. Fold into thickened Jello. Stir in nuts, drained pineapple and sliced bing cherries. Refrigerate 3 hours.

Waldorf Salad

8 diced apples
1 lg. can pineapple tidbits
1 1/2 c. mini marshmallows

1/2 c. walnuts
1 c. grapes
2 c. Cool Whip

DRESSING:
1 1/2 T. flour
3 eggs, beaten
1/2 c. sugar

1/2 c. water
2 T. margarine
1 tsp. vinegar

Cook dressing until thickened; let cool. Mix with the whipped cream. Toss in the drained pineapple, apples, marshmallows, walnuts and grapes. Chill 1 hour.

Chili Cheese Dip

1 lg. can Hormel chili
1 lb. Velveeta cheese, cubed

1 sm. jalapeño pepper, diced
1 lg. pkg. tortilla chips

Pour the can of chili in a glass bowl or microwave-safe bowl. Cube Velveeta cheese and mix in. Dice jalapeño pepper fine, making sure to remove the seeds for that is what makes it hot. Put into microwave and mix every 45 seconds until completely melted and incorporated.

Holiday Sugarless Eggnog

2 c. milk
2 T. cornstarch
3 1/2 tsp. Equal sweetener
2 eggs, beaten

2 tsp. vanilla
1/4 tsp. ground cinnamon
2 c. milk, chilled
1/8 tsp. ground nutmeg

Mix 2 cups milk, cornstarch and Equal in small saucepan; heat to a boil. Boil constantly for 1 minute. Beat eggs in medium bowl; mix 1/2 of the milk mixture into eggs. Add the egg mixture to remaining milk in saucepan. Cook over low heat until slightly thickened, 1 to 2 minutes, stirring constantly. Remove from heat and stir in vanilla and cinnamon. Cool to room temperature. Refrigerate until chilled. Stir in the other 2 cups chilled milk into custard mixture. Sprinkle with nutmeg.

For adult eggnog, stir in a little rum, to taste.

Blackberry Wine

1 gal. blackberries or any berry of
    your choice

4 gal. water
10 lb. sugar

Pit ingredients in a sealed basket. Put a small hose in the basket and out into a jar with a little water, so it can work and spew out into the water. After 4 to 6 weeks, put in jars and seal.

Homemade Root Beer

2 c. white sugar
1 gal. lukewarm water

5 T. root beer extract
1 tsp. dry yeast

Use some hot water to dissolve sugar. Mix other ingredients. Put into jars. Cover and set in sun for about 4 hours. Chill before serving. Ready to serve the next day. No need to bottle. Best served over ice cream.

Creamy Tapioca

2 qt. real milk
1 c. tapioca
5 eggs, separated
2 c. sugar

Pinch of salt
Vanilla
1/2 tsp. lemon extract

Slowly bring milk and tapioca to a boil. Add egg yolks, sugar, salt and vanilla to taste. Remove from heat. In a separate chilled bowl, beat egg whites until stiff. Add to milk mixture until well mixed. Chill.

Custard

5 c. hot milk
8 eggs
2 c. sugar

1/2 tsp. salt
2 tsp. vanilla

Heat milk. Pour over well-beaten eggs. Add other ingredients and mix well. Pour into glass dish or dishes. Sprinkle with nutmeg. Set in pan or pans of water to bake. Bake at 275° for about 3 hours.

Baked Butterscotch Pudding

4 T. butter
1/2 c. sugar
1 c. + 3 T. flour
1 1/2 tsp. baking powder
1/2 tsp. salt
3/4 c. cream

1/2 c. nuts
1 c. brown sugar
1/4 tsp. salt
1/2 T. maple syrup
1/2 tsp. vanilla

Mix the first 7 ingredients together. Pour into greased pan. Mix brown sugar and 1/4 teaspoon salt. Sprinkle on top of batter. Do not stir. Pour 1 1/4 cups boiling water over this. Bake at 350° for 40 to 45 minutes.
Delicious when eaten warm with ice cream.

Pine Scotch Pudding

1 c. flour
1 tsp. baking powder
1 tsp. baking soda
3 eggs
3/4 c. water

1 tsp. vanilla
1 c. crushed pineapple, drained
1 c. chopped nuts
1 tsp. almond extract

SAUCE:
1/4 c. butter
1 T. flour
1 c. brown sugar
1/4 c. water

1/4 c. pineapple juice
1 beaten egg
1 tsp. vanilla
1 tsp. almond extract

Beat eggs until fluffy; add sugar and vanilla. Fold into dry ingredients. Add pineapple and nuts. Bake at 350° until done.
Sauce: Melt butter in saucepan. Add flour, brown sugar, water and pineapple juice. Boil 3 minutes. Blend in egg and boil another minute. Add vanilla. Cut cake into small squares and toss with sauce and whipped cream.

Ozark Pudding

3/4 c. sugar
6 T. flour
1/4 tsp. salt
1 3/4 tsp. baking powder

1 c. finely-chopped dates
1 egg
1 tsp. vanilla
1/2 c. chopped nuts

Mix the 4 dry ingredients. Toss in finely-cut apples. Stir in egg, vanilla and nuts. Pour into 9-inch greased baking dish. Bake at 325° for 45 minutes. Serve with vanilla ice cream or whipped topping.

Cheesy Broccoli Soup

3 T. butter
3 T. flour
4 c. chicken broth
2 c. chopped broccoli
3/4 c. sliced carrots
1 sm. onion, diced
1 clove garlic

1/4 tsp. thyme
1/4 tsp. salt
1/4 tsp. pepper
1 c. heavy cream
1 egg
1 1/2 c. shredded Cheddar cheese

Melt butter; add flour and cook several minutes, stirring all the time. Remove from heat. Gradually blend in broth. Bring to a boil, stirring. Add next 8 ingredients. Cover and simmer 10 minutes, or until vegetables are tender. Blend cream and egg together; blend in several spoonfuls of soup mixture and return to vegetable mixture. Cook, stirring constantly, until thick. Blend in cheese. Serve when melted.

Ribbon Salad

2 pkg. lemon Jello
1 pkg. cream cheese
2 c. mini marshmallows
2 c. boiling water

2 c. whipped cream
1 c. salad dressing
2 c. crushed drained pineapple
1 c. grapes

Dissolve Jello in boiling water, then cool. When thickening, incorporate the rest of the ingredients. Refrigerate.

Navy Bean Soup

1 lb. navy beans	4 cubes beef broth
1 ham & bone	Carrots
2 sm. cans chicken broth	Onion

Soak beans overnight and drain off the next morning. Put back into pot and cover with water. Boil. After 20 minutes, add neck bones. In a separate pan, slice 6 to 7 carrots and 1 large onion. Bring to a boil until tender. Drain off water from the vegetables. Remove the neck bones from the other pot and pick the meat off; cut into chunks. Add vegetables and the remaining ingredients. Simmer on low until tender.

Homemade Mustard

Mustard powder (or mustard seed)   Vinegar

Make a paste with the ground mustard and vinegar. Add a little salt. If you have whole mustard seeds, soak them in the vinegar until they pop. Grind it up in a blender. Store in refrigerator.

Cappuccino Mix

1/2 c. instant coffee	1 c. powdered sugar
1/2 c. any flavor dry coffee creamer	2 c. powdered milk
3/4 c. Nestlé drink mix	

Mix all together and store in airtight container. Use 2 rounded teaspoons per cup of 8 ounces water.

Notes &
Recipes

Horse, Dog & Cat Treats

Baby calf isn't stupid

Lazy time at the farm

World War II.
The need of blood was
fierce. An American
soldier and a Portarican
girl. A perfect match.
A sister, and a family.
The young soldier gave his
blood so that the young
girl could live. Her
sister looked on as his
kindness grew on her.
Such terrible times yet
God smiled on them and
love grew to be a
romance that lived on
for over 61 years.

Horse, Dog & Cat Treats

Carrot Appaloooooosaaa's Horse Treats

1 1/2 c. sweet feed
2 c. bran
4 c. finely-shredded carrots

1 c. molasses
1/2 c. brown sugar
1 c. applesauce

Mix sweet feed, bran, carrots and brown sugar; mix well. Pour molasses and applesauce over the top. Add more bran if very sticky or more applesauce if too dry. Put a spoonful on the cookie sheet and flatten slightly. Bake at 300° for 1 hour. Flip over and bake an additional 45 minutes, or until dried out.

Annie Oatly's Soft Horse Cookies

1 1/2 c. oatmeal
3/4 c. sweet feed

3 c. bran
3/4 to 1 c. molasses

Mix well. Dough should be a bit thinner than play dough. Drop on greased cookie sheet. Bake at 350° for about10 minutes.
Watch them, as they have a tendency to burn quickly.

Belgian Bites Horse Treats

1 c. oatmeal
1 c. flour
1 c. finely-shredded carrots
2 T. sugar

2 T. corn oil
1/4 c. water
1/4 c. molasses
1 T. trace mineral salt

Mix all ingredients well. Use more flour if needed, or more molasses if too dry. Make small balls and place on greased cookie sheet. Spray top of cookies with Pam spray. Bake 15 minutes, or until done in a preheated oven of 350°.

Morgan Munchies Horse Cookies

1 c. oatmeal
1 c. bran
1 T. trace mineral salt

2 T. brown sugar
1/2 c. molasses

Mix all ingredients well. Roll into balls and place on greased cookie sheets. Bake at 350° for 8 minutes, or until outside is hard. Refrigerate uneaten cookies.

Paint Horse Peppermint Pats

1 c. oatmeal
1 c. bran
1 c. water

1 T. trace mineral salt
1 T. peppermint extract

Mix all ingredients well. Add more or less of the bran or molasses, if wet or dry. Make into balls and place on greased cookie sheet. Bake at 350° for 8 to 10 minutes. The outsides should be crunchy. Refrigerate uneaten cookies.

Bay Bars Horse Treats

2 c. oatmeal
1/2 c. finely-shredded carrots

4 T. molasses
1/2 c. brown sugar

Mix well and add enough water to incorporate ingredients. Make into a soft dough. Place spoonfuls on greased cookie sheet and bake at 350° until done, 10 minutes or so.

Carrot Cruncher's Horse Cookie

1 c. oatmeal
1 c. flour
1 c. shredded carrots
2 T. corn oil

1 T. sugar
1 T. mineral oil
1/4 c. molasses

Mix ingredients well. Add more flour, if needed. Add more molasses, if needed, to make dough. Roll into balls and place on greased cookie sheet. Bake at 350° for 15 minutes, or until done.

Stallion Sweeties Horse Cookies

2 c. oatmeal
3/4 c. sweet feed

3 c. bran
1 c. molasses

Mix ingredients well. Place on greased cookie sheet. Bake at 375° for 8 minutes, or until done.

Donkey Delights Horse Cookies

1 c. oatmeal
1 c. flour
1 c. shredded carrots
1 T. trace mineral salt

2 T. sugar
2 T. corn oil
1/4 c. water
1/4 c. molasses

Mix well. If dough is too dry, add more water; if too moist, add more oatmeal. Roll into balls and press a bit on the greased cookie sheet. Bake at 350° for 15 to 20 minutes.

Clydesdale Cut-Outs, Horse Cookies

4 c. bran
4 c. applesauce

1 T. trace mineral salt

Mix ingredients well. If the dough is too wet, add more bran; if it is too dry, add applesauce. Roll dough out to 1/4 of an inch. Cut with cookie cutter horse shapes, boots, etc. Place on cookie sheet and allow to air dry.

Nicker Nibbles Horse Cookies

2 c. flour
5 c. oatmeal
1/2 c. corn oil

1 clove garlic, crushed
1 c. shredded carrots

Mix all ingredients well. Roll into small balls and flatten slightly. Put on microwave-safe plate and bake on HIGH for 6 minutes per batch.

Whinnies Horse Cookies

2 sticks butter
3/4 c. brown sugar
2 eggs
1 c. raisins

2 c. oatmeal
1 c. alfalfa pellets
2 T. trace mineral salt

Beat eggs, butter and sugar. Add oatmeal, trace mineral and raisins. Mix in alfalfa pellets. Place spoonfuls on greased cookie sheets and bake at 350° for 8 to 10 minutes.

Electrolyte Cookies for Horses

2 c. oatmeal
3/4 c. sweet feed
3 c. bran
1 c. molasses

1 c. apple juice
12 to 25 oz. electrolyte powder,
 depending on how strong you want
 them

Mix all ingredients well. If too dry, add more apple juice; if too wet, add more oats. Place spoonfuls on greased cookie sheet and bake at 325° for 35 to 40 minutes. Be careful as this cookie has a tendency to burn.

Homemade Fly Spray for Horses

2 T. liquid soap
2 c. white wine vinegar

1 c. water

ESSENTIAL OILS:
1 c. citronella oil
1 tsp. eucalyptus oil
1 tsp. lavender oil

1 tsp. sandlewood oil
1 tsp. tea tree oil

In large spray bottle, add vinegar, water and citronella oil. In the liquid soap, add all of the essential oils. Pour the liquid soap/oil mixture into the spray bottle, being sure to scrape all in. Shake before using. Test in small area on horse to make sure it is not allergic.

Fly Spray For Horses

2 lg. lemons

Rosemary sprigs

Slice lemons thin and put in pot. Cover with boiling water. Place 4 sprigs of rosemary in. Bring to a boil and boil for 10 minutes. Cover and let steep overnight. Strain and pour into spray bottle. Test on small spot of horse to make sure it is not allergic.

Internal Horse Fly Repellant

1/4 c. cider vinegar

Add cider vinegar to your horse's 5-gallon bucket. It raises the PH of the horse's blood sufficiently to stop flies from biting.

The idea of the vinegar was written in a Civil War diary. After research, it was found that it raises the PH of the animal's blood.

Horse Fly Spray Italian Style

1 c. crushed garlic   5 c. water

Crush garlic and put in saucepan. Cover with boiling water and cover. Steep overnight. Strain off and put in spray bottle. Test on horse in small area to make sure they are not allergic. I keep the garlic that I strained off and give 2 tablespoons a day to the horse with its feed. It helps to guard off mosquitoes if taken internally. Please keep strained garlic in refrigerator.

Bow Wow Breath Refreshers

2 c. brown rice flour   1/2 c. fresh mint
1 1/2 T. activated charcoal, not   1/2 c. fresh parsley
   briquettes (ask drugstore)   2/3 c. milk
3 T. canola oil   Substitute soy milk, if needed
1 egg

Bruise and finely chop mint and parsley. In a separate bowl, mix remaining ingredients. Incorporate both and mix well. Place small spoonfuls on greased cookie sheet. Bake 15 to 20 minutes. Cool and store in refrigerator.

Bow Wow Bacon Bites

2 1/2 c. whole wheat flour   8 T. bacon grease
1/2 c. powdered milk   1 egg
1 tsp. brown sugar   1/2 c. very cold water

Mix all ingredients well. Roll out dough to 1/4-inch thick. Cut into shapes. Poke each cookie with a fork before putting in the oven at 350° for 25 to 30 minutes. Store in an airtight container.

Softies Dog Cookies

2 1/2 to 3 jars chicken or beef baby   1/4 c. dry milk powder
   food   1/2 c. Cream of Wheat

Mix all ingredients well. If too dry, add a little broth. Dough should be firm enough to form a ball. Roll into small balls and place on a well-greased cookie sheet. Flatten with a fork. Bake at 350° for 15 minutes, or until brown. Cool on wire rack. Refrigerate uneaten cookies.

This cookie freezes very well, and they like them frozen for a cool treat in the summer.

Milky Bones Dog Bones

3/4 c. hot water
1/3 c. margarine
1/2 c. powdered milk

1/2 tsp. salt
1 egg
1 to 1 1/2 c. whole wheat flour

Beat egg well. Pour hot water over the margarine. Stir in powdered milk. Add salt and well-beaten egg. Add flour a little at a time. Knead for a few minutes until dough is firm. If sticky, add more flour; if too dry, add a bit more water. Roll dough out to 1/2-inch. Cut in bone shapes and bake at 325° for 50 minutes. Leave out until dries hard. Store in Ziploc bag.

Goodboy Biscuits Dog Bones

3 to 4 c. whole wheat flour
2 c. oatmeal
1 c. milk

1/2 c. broth
2 beef or chicken bouillon cubes
1/2 c. meat drippings

Dissolve bouillon in broth. Add powdered milk and grease drippings. In a separate bowl, mix flour and oatmeal. Pour the hot liquid over the top and mix well. Press onto an ungreased cookie sheet. Cut into shapes on the cookie sheet. Bake at 300° for about 1 hour. Turn the oven off and allow the cookies to remain in oven until cooled. This helps to harden them. Refrigerate uneaten cookies.

Nutty Buddy Bones Dog Cookies

1 3/4 c. flour
1/2 c. sesame seeds
1/2 c. brown sugar
14 T. butter

1/2 c. ground walnuts
1 egg
1/2 tsp. vanilla
2 T. toasted wheat germ

Combine butter and sugar; add egg and beat well. Add vanilla, and then add the dry ingredients gradually. Mix into a stiff dough. Roll out to 1/2-inch thick. Cut out with cookie cutters and place on an ungreased cookie sheet. Bake 12 to 15 minutes.

Keeps very well in an airtight container for 2 weeks.

Liver Chomps Dog Bites

1 lb. puréed liver
2 c. cornmeal

1 c. flour
1 tsp. garlic salt

Mix all ingredients well. Pour out on greased cookie sheet and spread well. Make a thin layer. Bake at 350° for 20 minutes. Cool and cut into pieces. Store in an airtight container. Keep uneaten portion in refrigerator.

Liver Treats Dog Cookies

2 jars liver or beef baby food   1 c. powdered milk
1/3 c. wheat germ

Mix all ingredients well; add more wheat germ if too wet, or broth if too dry. Drop spoonfuls onto greased cookie sheet. Bake at 350° for 12 to 15 minutes, or until done. Cool and refrigerate.

Barko Bites Dog Cookies

8 slices cooked, crumbled bacon   1/3 c. powdered milk
4 eggs, beaten   2 c. wheat flour
1/8 c. bacon grease   2 c. wheat germ
1 c. water   1/2 to 1 c. cornmeal

Mix all and drop spoonfuls onto greased cookie sheet. Bake at 350° for about 15 minutes. Turn off the oven and allow cookies to cool in oven. Leave cookies overnight to harden. Store in airtight container and refrigerate unused portions.

Itty Bitty Bones

2 eggs   4 T. cold broth
3 T. rice flour   2 c. whole wheat flour
2 T. wheat germ   2 to 4 T. powdered milk
1/2 tsp. salt

Beat eggs well; add cold broth. Gradually combine dry ingredients into the egg mixture. Pat out to about 1/2-inch thick. Cut with cookie cutters. Bake at 350° for 25 minutes. Flip over and bake 25 more minutes. Cool on racks. Store in an airtight container and store uneaten portions in the refrigerator.

Our little rat terrier, Itty Bitty, loves them.

Peanut Budder's Dog Cookies

2 c. whole wheat flour   1 c. peanut butter
1 T. baking powder   1 c. milk

Mix all ingredients well. If too wet, add more flour; if too dry, add more milk. Roll out dough and cut into shapes. I roll this dough about 1/4-inch thick. Bake at 375° on greased cookie sheet for about 20 minutes, but be careful because this cookie burns easily, so keep a watchful eye.

Peanut Butter Gourmet Dog Cookies

1 1/2 c. water
1/2 c. oil
3 eggs
3 T. peanut butter

1 tsp. vanilla
2 to 2 1/2 c. flour
1/2 c. cornmeal
1/2 c. instant oats

Beat eggs, vanilla, oil and peanut butter well. Combine dry ingredients and mix well. Add more or less flour and water to make a dough stiff enough to roll out. Knead dough for a bit. Roll out to 1/2-inch thick and cut with cookie cutters. Place on greased cookie sheet and bake in a preheated oven at 400° for 20 minutes. Leave in oven to cool to a crisp. Store in an airtight container in refrigerator.

Cheesy Dog Crunchies

1 c. flour
1/2 c. milk
3 T. peanut butter
1/2 c. Parmesan cheese

1 tsp. baking powder
1 egg white
1 to 3 T. chicken broth

Mix all ingredients well and beat for 1 minute. This dough should be the consistency of pancake batter. Pour 2-inch drops onto a greased cookie sheet. Bake at 400° for 15 to 20 minutes, or until brown. Store in airtight container in the refrigerator.

Gourmet Dog Biscuits

1 lb. raw liver, chopped
1 1/2 lb. flour
1 to 1 1/2 c. quick oats
4 bouillon cubes

Approx. 1 c. hot water
2 eggs
2 T. meat drippings

Chop liver or purée in food processor. Dissolve bouillon in hot water. Beat 2 eggs very well. Mix these ingredients well. Gradually combine flour and oats. Add more or less to make a dough to roll out. Pat out to 1/2-inch thick. Cut with cookie cutters and place on greased cookie sheets at 450° for 1 hour. Refrigerate uneaten portions. Will keep for 2 weeks.

Tuna Treats For Cats

1/2 can tuna
1 c. bread crumbs

1 egg
1 T. oil

Beat egg and oil until incorporated. Mash the tuna and combine with bread crumbs. Combine egg, tuna and crumbs together and mix well. Drop 1/4 teaspoon onto ungreased cookie sheet. Bake at 350° for 8 minutes. Store in airtight container in the refrigerator.

Catnip Crunchies

1 to 1 1/2 c. whole wheat flour
1/4 c. rice flour
2 tsp. catnip
1 egg
1/3 c. milk

2 T. wheat germ
1/3 c. powdered milk
1 T. molasses
2 T. butter

Beat egg, butter and molasses well. Combine the remaining ingredients. Dough should be able to be rolled out. Of too dry, add more milk, and if too wet, add more flour. Roll out to 1/4-inch thick. Cut into small pieces and place on greased cookie sheet. Bake at 350° for about 20 minutes. Refrigerate in airtight container.

Chicken Liver Treats For Cats

1 lb. chicken livers
1 1/2 c. cornmeal
2 eggs

1/2 c. powdered milk
2 T. molasses
1 lg. garlic clove, chopped

Put all ingredients in a food processor, liquid first and incorporate dry. Pour in a greased 9x13-inch baking pan. Bake at 400° until the sides pull away from the pan. Insert knife in the middle and if it comes out clean, it is done. Cool and cut into bite-size pieces. Store in an airtight container in the refrigerator.

Notes &
Recipes

Soaps, Salts, Fly Sprays
& Miscellaneous

Barnyard Days

Baryard Days

My name is Patuty. I am a Boer goat. I came to the attention of Airocolina and my life changed for the better. I was at the sale barn, where she always goes to buy or pick up the sick and injured animals. The only way I could walk was on my elbows. My skull was cracked and my horn able to move, which caused me great pain.

Even though I want to tell her how it happened, I can't. She took me home and put me in my own warm clean, safe stall. I was frightened at first. I could hear the other goat talking to me, telling me it was ok. She came into me and gave me some warm mash. It felt so good to put something warm in my tummy. She checked on me all through the night. I could tell that she was a kind and loving person.

My skull has healed and I feel like I have found my place. I will help her all I can with her Barnyard Days for the senior home and the daycare centers. I can let a few hands pet me and give me grain. I feel like I am paying her back for letting us stay here and for her kindness.

Patuty

Soaps, Salts, Fly Sprays & Miscellaneous

Flea Collars For Cats and Dogs

Rosemary   **Oregano**

Cut a strip of cloth wide enough to fold over and stitch one end. Stitch along the side. Make a 50/50 mix of the rosemary and oregano. Seal the other end. Now stitch Velcro on each end and put around dog's neck.

Natural Insecticide

Basil   **Basil plants**

If you have a problem with any type of flying insects, try keeping a basil plant or two around the house. Keep the plant well watered from the bottom, this will cause the plant to release additional aroma. Hanging small muslin bags with fresh dried basil will also repel flying insects. Works with anything with wings.

Vitamins and Minerals

Vitamins and minerals are very important to your pet's health. Save the water from boiled vegetables or liquid from a crock-pot and mix it with your animal's food for additional nutrients.

Buttermilk Bath

1/2 c. dry buttermilk or dry milk   2 T. cornstarch
1/2 c. nonfat dry milk   1/4 tsp. of any essential oil

Mix ingredients together with a wire whisk. Pour into airtight jar or resealable plastic bag. Pour 1/4 cup under running water.

Herbal Milk Bath

1 c. cornstarch
1 c. dry milk powder

2 tsp. your favorite herb

Combine all ingredients in your food processor. Add herbs and blend.

Do-It-Yourself Bath Salts

1 c. Epsom salts
1 c. sea salt

1/2 c. baking soda
1/2 tsp. essential oils

Combine all ingredients with a whisk. If your salt is large crystals type, grind your salt first in a food processor until finely grained.

Bubble Milk Bath

1 c. powdered milk or buttermilk
  powder
1/2 c. oatmeal
5 T. cornstarch

2 T. cream of tartar
1/4 c. powder bubble bath
1/2 tsp. essential oil

Food process oatmeal until powdered. Mix in powdered bubble bath. Now add remaining ingredients. Add fragrance and process until powdery. Store in glass jar. Use 1/4 cup per bath under running water.

Cleansing Body Polish

1 c. fine seal salt
1/4 c. jojoba oil or olive oil

1/2 c. handcrafted soap
1/2 tsp. essential oil

Combine all ingredients in bowl and mix well. Store in small plastic jar that you can keep by the shower.

Directions for Use: Use in shower on dry skin (otherwise the salt will melt instead of scrub). Rub and scrub anywhere you need to exfoliate and moisturize.

Bath Bombs

4 T. citric acid
4 T. cornstarch
1/2 c. baking soda

3 T. Monoide Tahiti oil or coconut oil
or almond oil
1/2 tsp. essential oil

Sift the first 3 ingredients in a bowl; mix well with wire whisk. Slowly drizzle oils over the top, mixing well. Now add the fragrance. Take 1 tablespoons of mixture and shape into balls or press in large chocolate mold. Freeze until hard, then pop out. Place on waxed paper for 3 to 4 hours, or longer. Gently reshape if needed.

Goat's Milk Soap

3 pt. ice cold goat's milk
1 (12 oz.) can Red Devil lye
5 to 6 lb. lard

3 oz. glycerin
2 T. Borax
1/3 c. honey

Using a stainless steel pot, slowly pour the lye into the ice cold milk, stirring constantly with a wooden spoon. The milk will heat up very quickly. If you add the lye too fast, the milk will scorch and curdle. The milk will turn an orange color and curdle a little bit, but don't worry. Add the honey and let cool down to 85°.

While the milk is cooling, warm the lard to 90° and slowly pour the lard into the lye/milk, stirring constantly. Now add the glycerin and Borax. Beat soap with mixer to trace. When the mixture starts to trace (thicken to a thin pudding), in 10 to 20 minutes, pour the mixture into molds, garbage bag-lined boxes, short, long boxes like cases of 6-pack pop comes in. Leave undisturbed overnight. Cut the bars using fishing line. Stack the bars on a cookie sheet lined with a large paper bag. Let soap cure 6 weeks.

Homemade Liquid Soap

4 1/2 gal. soft water
1 can lye
7 c. melted grease (1/2 tallow, 1/2 lard preferred)

1 c. ammonia
2 c. Borax
3 1/2 c. Tide liquid soap

Fill crock or plastic bucket half full of water. Add remaining ingredients and mix well. Fill with water to make 5 gallons. Stir a few minutes several times a day for about a week.

Powdered Homemade Soap

1 can lye
3/4 c. Borax

3 qt. cold water
4 1/2 lb. melted fat, 1/2 tallow

Dissolve lye in cold water; add Borax. When dissolved, add melted fat. Stir 15 minutes, then stir frequently for 24 to 36 hours.

Cinnamon Soap

4 oz. unscented glycerin soap
10 drops cinnamon oil

1 drop food coloring (opt.)

Heat glycerin and remove from the heat. Add cinnamon oil and food coloring. Pour into mold and let cure for 4 to 5 hours. Remove and cut and let air cure about 1 to 2 days, then wrap.

Liquid Gel Soap

2 c. grated soap bar flakes
1/2 gal. soft water

3 T. glycerin

Mix in large pot or Dutch oven over low heat, stirring occasionally, until soap is dissolved. Transfer to a jar and cover tightly. For a thinner soap, use 1 gallon of water.

Vanilla Almond Soap

2/3 c. whole almonds, ground
1 (14 oz.) bar castile soap, grated
1/4 to 1/2 c. distilled water

1 T. almond oil
1/8 tsp. vanilla fragrance

Grind almonds into a powder, using a food processor or a coffee grinder. Grate soap and set aside. In a large saucepan, heat water to a boil and reduce heat to a simmer. Remove from heat and add grated soap. Mix until dissolved. Add the almond and vanilla oils and stir until blended. Pour into soap mold and let stand 5 hours. Pop out of mold and let air cure for 2 days before wrapping.

Coconut Olive Soap

1 c. olive oil	8 oz. olive oil
1 1/4 c. coconut oil	12 oz. coconut oil
1 c. melted tallow (animal fat)	8 oz. tallow
2 heaping T. lye (Red Devil)	3.4 oz. lye
1/2 c. soft cold water	9 oz. water
OR REVISED TO:	

Combine cold water and lye. Be very careful not to splash on skin or eyes. If you do, rinse thoroughly with water. Mixing these two will heat up. Allow it to cool to 100° to 125°; it is best to let it do this outside.

Combine oils and gently heat (do not burn) and melt tallow to about 100° to 125°. Combine lye solution and oils and tallow that were previously melted. Be careful not to splash. Stir until the soap "traces". This is the process where it slightly thickens like custard, not the instant type, but cooked custard. This process takes about 15 minutes. If this hasn't occurred, then stir every 5 minutes for the next 15 minutes. Drop a little in the sink and if it supports itself, it is ready. Pour into shallow plastic bag-lined cardboard box.

After a few days, soap can be turned out. Let cure for a few days more and then cut it in pieces. Let cure a few more days until firmer and then wrap.

Peaches and Cream Soap

1 (4 oz.) bar Castile or Ivory soap	1 T. sweet almond oil
1/4 c. distilled water	1/8 tsp. peach fragrance
1/4 c. powdered milk	1 drop orange food coloring

Grate soap and set aside.

In a saucepan, heat water. Stir in grated soap until dissolved. This will form a sticky mass. Remove from heat. Add powdered milk and oils. Spoon into mold and let set up about 4 to 5 hours. Remove from mold and let air cure 1 day; wrap.

Note: You can substitute strawberry oil, etc.

Mechanic's Hand Cleaner

2 bars Lava soap, ground	3 T. turpentine
1 c. powdered Borax	1 T. sweet orange essential oil

Take 1 cup of your grated Lava soap and set aside.

Now work your turpentine and oil into the Borax. Work it in with your fingers until there are no lumps. Mix into your soap that is set aside. Store in wide-mouth jar so that he can get it open easily with greasy hands.

Play Dough

1/2 c. salt
1 c. flour
1 c. water

2 tsp. cream of tartar
1 T. vegetable oil
Food coloring

Mix salt, flour and cream of tartar in a large kettle; add water and vegetable oil. Stir until well blended. Stir in a few drops of food coloring. Cook over low heat until mixture gets rubbery, stirring constantly. Turn dough onto table and knead with palm of hand until pliable and easy to work. Keep in a baggie or covered container.

Bath Salts

2 lb. Epsom salts
4 T. mineral oil

2 T. liquid glycerin
Food coloring of choice

Pour Epsom salts into large bowl. Drizzle oil and glycerin over the top and mix in well. This will take some time; press into it until not clumpy. Add food coloring and start the process over. Add scented oil of your choice and mix well. Store in decorative container in bathroom. Pour 1 cup into tub under running water. Relax and enjoy.

Aromatic Herbal Bath Salts

2 lb. Epsom salts
4 T. glycerin
4 T. liquid bath soap

2 tsp. lavender buds or rose petals,
    dried & crushed (be creative & use
    your imagination)

Mix the first 3 ingredients. Mix well; when it is incorporated, add your dried buds. Pour under running water, step into, relax and enjoy.

Bath Tea Bags

Lavender buds
Vanilla beans, chopped

Cheesecloth
Juet string

Cut your cheesecloth into a 4x4-inch square. Place 1 tablespoon lavender buds and small piece of vanilla bean into center of cloth. Draw up the sides and tie with piece of string. Drop bag into your tub and let the aroma set you free.

Rose Hip Tea Bath Bags

2 c. rose hips   Juet string
4"x4" pieces of cheesecloth

Place 1 tablespoon rose hips in center of square and draw up the edges. Tie with piece of string. Drop in running bath and enjoy.

Mint Bath Tea Bags

Garden-fresh mint, cut & chopped   Juet string
4"x4" pieces of cheesecloth

Place 2 tablespoons chopped fresh mint into center of piece of cheesecloth. Draw up edges and tie with string. Drop into your bath.
I love the smell of mint; it is very invigorating.

Mildew Remover

2/3 c. Tide with Bleach   1 1/2 qt. bleach
2/3 c. Spic & Span   1 gal. water

Mix together and spray mixture on. Wait 20 minutes and then rinse with hose.

Ice Pack

1 c. alcohol   2 c. water

Mix and pour into double ziplock quart-size bags, one inside the other, and freeze. This is the jelled kind of ice pack that does not freeze hard. Do not eat this.

Little Tykes Upset Tummies

6 oz. 7-Up   3 oz. orange juice
4 oz. Pedialyte   1 tsp. sugar

Give 1 ounce every 20 minutes for 1 1/2 hours, then 3 to 5 ounces every 2 hours. Give no milk until green stool is gone.
I got this from a baby specialist, and it sure helped with ours.

Lavender Baby Wipes

2 c. boiling water   **3 T. lavender baby bath**
2 T. lavender baby oil

You will need 1 roll of paper towels. Put oil, baby bath and water into container. Remove cardboard roll from paper towels; place towels into mixture and cover.

Sterling

The trailer doors opened and standing in the front on the trailer, dripping with sweat, was a little, silver, frightened pony. I reached up to grab the center bar to step into the trailer. The whites of his eyes widened. He seemed to become more frightened by the second. I stepped up and he went nuts. I jumped out of the trailer and he calmed back down.

I looked at him from outside and could see that he was badly beaten. Open lesions and cuts were all about his body. He had two split welts on his upper rump on almost his back. He had been, what you would call, caned, where someone slaps the skin with a cane so hard that it splits it wide open.

You could tell he was not going to be easy. I called our vet and she came over. We dart-gun tranquilized him. With the help of my farm hands, we were able to put him on a large cart and get him down to the barn. While he was out, the vet worked on the puss-filled lesions. She then turned her attentions to his back. He had been beaten so bad that he looked like a hunch back.

For the next few weeks, we had to introduce him to grain and hay of good quality. A sample of his manure showed that he had been eating his own manure. We wormed him.

I slide his stall door open and his eyes were large and fixed on me. I took one step into his stall and he ran to the opposite corner. I tried to pick the stall with him in it, and he went nuts. It has taken up to 3 years of tender, loving care to be able to approach him and pet him. He doesn't like shovels, pitch forks, or anything with a handle.

We named him Sterling and this is his story. He continues to live at our shelter and has therapy every day, and will continue to do so for the rest of his life.

A Beautiful Spring Day

It was a beautiful, sunny spring day at the feed store. The sounds of the day-old chicks peeping in the background. She was standing on a step stool hanging up the "chicks for sale" signs. It was 1951, Easter was right around the corner. He walked in to buy his feed and get his free chicks. He glanced up at her on the step stool and reached up to help her step down. When their eyes met and their fingers touched, it started a life-long romance that lasted until their deaths at ages 82 and 70.

Bon Bons the Goat

It was about 88 degrees that day. Sunny, the smell of fresh mowed alfalfa was in the air. I saw a van coming down my lane. I walked toward the drive. I could see that it was an animal cruelty van. I wondered what we could be getting into.

I greeted the female officer. She said we have a bad one for you. We walked around to the back of the van and I could hear that it was a goat. She opened the door, and here stood an Alpine goat. She looked ok until the officer explained that she had been a sex toy for a very sick man.

I called for our farm hand. He came with a lead rope and started to walk her down to the barn. I could see the blood on her and the tie marks on her rear hooves.

Bon Bons healed, but will never be able to get pregnant. She has different ideas toward the other goats. Most of the time she has free roam around the shelter. She is sweet and loving. We were blessed the day she came to stay.

Notes &
Recipes

Index

Breads & Rolls

Camp Fire & Open Fire Recipes

Sauces, Soups, Salads, Puddings, Beverages & Jams

Horse, Dog & Cat Treats

98

Soaps, Salts, Fly Sprays & Miscellaneous

About the Author

I have studied under five vets for marsupials and exotics. I have a Shelter and Rescue called Settler's Pond.

I am a not for profit and write books to supplement my shelter. I have had the shelter for about 10 years. I like to write books about the animals in my shelter and how they came to be. I took care of my bed-ridden mother in law for 10 years until her death in 2005. I hope to continue my research on macro pods and other exotics and writing my books help the shelter to continue. You can visit my web site www.settlerspondshelter.net for more info and further Bio.

978-0-595-48965-7
0-595-48965-6